Lecture Notes in Economics and Mathematical Systems 590

Sobei Hidenori Oda
(Editor)

Developments on Experimental Economics

New Approaches to Solving Real-world Problems

With 64 Figures and 36 Tables

Springer

Professor Sobei Hidenori Oda
Faculty of Economics
Kyoto Sangyo University
Motoyama, Kamigamo, Kita-ku
Kyoto 603-8555
Japan
oda@cc.kyoto-su.ac.jp

Library of Congress Control Number: 2007927160

ISSN 0075-8442

ISBN 978-3-540-68659-0 Springer Berlin Heidelberg New York

Springer is a part of Springer Science+Business Media

springer.com

Production: LE-TeX Jelonek, Schmidt & Vöckler GbR, Leipzig
Cover-design: WMX Design GmbH, Heidelberg

SPIN 11954422 88/3180YL - 5 4 3 2 1 0 Printed on acid-free paper

Preface

This volume presents papers and speeches given in *the Experimental Economics Week in Honour of Dr Vernon L. Smith* held in Okayama and Kyoto, 13-17 December 2004, which consisted of Dr Smith's public speech and *the International Conference on Experiments in Economic Sciences: New Approaches to Solving Real-world Problems.*

Despite having a short history, experiments are now considered indispensable in economics as in other fields of science and engineering. As Dr Smith's Nobel Prize (2002) shows, experimental economics has now established itself in modern economics. In such an environment, researchers are expected to develop the tradition with new ideas in new fields for solving various problems in the real world. The Experimental Economics Week, which was organised to explore new fields for experiments with new approaches, provided a unique opportunity for those who were engaged or interested in experiments in their fields to discuss experimental approaches from various standpoints.

Economic experiments broaden and deepen our understanding of human behaviour, the economy and their interdependence. Some experiments are designed to observe how people behave. Experimenters control subjects' economic environment to guess their strategies, which are not always apparent in the real world. The environment may be game-theoretic (a person's gain or loss is affected by other persons' actions) or non-game-theoretic. In either case what is checked is subjects' behaviour. Some experiments are done to see how market or other economic systems work. In such experiments, subjects are not checked by the game but check the game for the experimenter to see the performance or the dynamics of the system the game represents. Some experiments examine how individuals' behaviour affects and is affected by the whole system. In the conference of the Experimental Week, the keynote and invited speakers taught important lessons about what economic experiments can discover and how they can contribute to the real world, while researchers from various disciplines presented various experimental works and applications in parallel sessions. The reader will find the fruits of this week in the following pages.

Part One provides Dr Smith's public speech and his keynote speech for the conference. The reader will find his insight and vision about the history of economics and the future of experimental economics. Part Two contains papers by seven of the invited speakers of the conference. The reader will find new ideas of the leading researchers in the field of experimental economics. The remaining parts provide twenty-one papers selected from the presentations in the parallel sessions of the conference. For the sake of the reader's convenience, the papers are divided into four according to the topic of each paper: Non-game theoretic decision making, Game theoretic decision making, Performance of Systems, and Interdependence of System's performance and individual behaviour.

The papers cover a broad range: experimental economics, experimental management theory, experimental accounting, computational economics, social engineering, etc. I hope the reader will enjoy and use the ideas in the book to advance our understanding and improve the real world.

The Experimental Economics Week in Honour of Dr Vernon Smith was sponsored by Kyoto Sangyo University (KSU). The international conference of the Week, namely *International Conference of Experiments in Economic Sciences: New Approaches to Solving Real-world Problems (EES2004)*, was organised and sponsored by KSU and the Hayashibara Foundation in Okayama. It is also an activity of the Open Research Centre Project *Experimental Economics: A new method of teaching economics and the research on its impact on society (2001-2005)*. The sessions of experimental accountings are supported with the cooperation of Research Institute for Economics and Business Administration, Kobe University, while the sessions of co-creative decision making are supported with the cooperation of Research into Artifacts Center for Engineering, The University of Tokyo. I should like to thank The Ministry of Education, Culture, Sports, Science and Technology and the above-mentioned organisations. I should like to extend my thanks to the contributors of the papers, the participants of the conference, the audience of the public speech and those who worked for the conference with me as the member of the organising committee of EES 2004 : Prof. Fumihiko Goto, Prof. Katsuhiko Nagase, Prof. Akira Namatame, Prof. Kanji Ueda, Prof. Hidetoshi Yamaji and Prof. Yoshio Iida. I should also like to thank Mrs Barbara Fess, the editor of Springer Verlag, who has shown a great deal of patience in seeing this book through the press. Last, not at the least, I should like to thank my wife Hatsuko and the young researchers and graduate students who studied with me and now are engaged in the Open Research Centre Project *Experimental Economics: Who learns what from economic experiments? (2006-2008)*.

April, 2007 *Sobei H. Oda*

Contents

Part III Non Game-Theoretic Individual Decision Making

Part IV Game-Theoretic Individual Decision Making

Part V Performance of the System

List of Contributors

Vernon L. Smith
George Mason University

Tatsuyoshi Saijo
Osaka University

James C. Cox
Georgia State University

Ido Erev
Technion -Israel Institute of Technology-

Daniel Friedman
University of California, Santa Cruz

Robin M. Hogarth
Universitat Pompeu Fabra

Shyam Sunder
Yale University

Amy Farmer
University of Arkansas

Balázs Adamcsel
Loránd Eötvös University

Cary Deck
University of Arkansas

Chairs Kaskiris
University of California, Berkeley

Christiane Schiwieren
Universitat Pompeu Fabra

Dao-Zhi Zeng
Kagawa University

Daan van Soest
University of Amsterdam

Eric O'N. Fisher
California Polytechnic State University, San Luis Obispo

Fei Song
Ryerson University

Gen Masumoto
Kyoto Sangyo University

Hiroki Nakayama
The University of Tokyo

Hiroyasu Yoneda
Kyoto Sangyo University

Jane Vyrastekova
Tilburg University

Jordi Brandts
Institut d'Anàlisi Econòmica

Juergen Huber
Universitaet Innsbruck

Kanji Ueda
The University of Tokyo

Kazuhito Ogawa
Hiroshima City University

Kouhei Iyori
Hiroshima Shudo University

László Gulyás
Loránd Eötvös University

Matthias Sutter
Universitaet Innsbruck

Metin Senbil
Hiroshima University

Michael Kirchler
Universitaet Innsbruck

Nariaki Nishino
The University of Tokyo

Nobutada Fujii
The University of Tokyo

Pravin Varaiya
University of California, Berkeley

Rahul Jain
University of California, Berkeley

Ram Rajagopal
University of California, Berkeley

Ryoko Wada
Keiai University

Ryuichi Kitamura
Kyoto University

Shunichiro Sasaki
Osaka University

Sobei H. Oda
Kyoto Sangyo University

Susumu Fujii
Kobe University

Takemi Fujikawa
University of Western Sydney

Tatsuhiko Kato
Meiji University

Tomomi Kito
The University of Tokyo

Toshiya Kaihara
Kobe University

Utteeyo Dasgupta
University of Arizona

Yasuyo Hamaguchi
Nagoya City University

Yoshio Iida
Kyoto Sangyo University

Part I

Vernon L. Smith's Speeches

Public Speech: "Markets, Capital Markets and Globalization"

Vernon L. Smith

George Mason University

I want to begin with a quotation from David Hume on trade. He was writing in the 18th century in Scotland. David Hume was part of what we call the Scottish Enlightenment, and the two most important figures in the Scottish Enlightenment were David Hume and Adam Smith. And this is David Hume on trade:

> Manufacturers gradually shift their places leaving those countries and provinces which they have already enriched and plying to others, whither they are allured by the cheapness of provisions and labor, till they have enriched these also and are again banished by the same causes.

My message today is an optimistic message about the future, about economic betterment and the development of world trade and world resources. It is about exchange and markets, without which people cannot engage in task and knowledge specialization. It is this specialization that is the secret of all wealth creation. There is no other source of sustainable human betterment. We all function simultaneously in more than one world of exchange. Those worlds overlap, as we live first in a world of personal exchange, trading favors and friendship and of building reputations based on trust and trustworthiness in small groups and families; and, then secondly we live in a world of impersonal exchange, where communication and cooperation gradually emerged in trade with strangers, through markets. In this talk I want to speak of two kinds of markets. First, markets for commodities and services. These are the foundation of wealth creation. And secondly, I want to speak of markets for capital, or stock markets. Capital and stock markets are far more volatile and more unpredictable than are the existing commodity and service markets, but their function is to anticipate the commodities and services of the future.

I will also discuss globalization which is really nothing more than a new word for an ancient process of migration and development that began a long time ago when our common ancestors first walked out of Africa.

Exchange has its origin in reciprocity and sharing norms in the family, the extended family and tribes. This personal exchange allowed task specialization in hunting and gathering that laid the basis for enhanced productivity and welfare, which in turn enabled early peoples to migrate all over the world. Globalization for us all began when our ancestors moved out of Africa over 50,000 years ago, settled the Iberian peninsula and Southern Europe, Asia, then Australia somewhere around 50-40,000 years ago. These people discovered America, probably about 13-12,000 years ago, although it might have been earlier. And then finally they settled the islands of New Zealand and Madagascar only 1,000 years ago. So, long before the square rigger sailing ship, our ancestors had settled every continent, except Antarctica, and all of the major islands.

Early peoples, before nation states, traded tools, weapons, and public goods like symbols, customs, crests and unmolested rights of access to trade routes and hunting grounds.

In the laboratory, we believe we see the ancient norm of reciprocity and trading favors as it emerges in two person games between anonymously matched subjects, many of whom use trust and trustworthiness to achieve cooperative outcomes that consciously maximize their joint benefits. They voluntarily avoid choosing outcomes that take for themselves without giving something in return to their matched counterpart.

But when the same subjects, who consciously cooperate for betterment in elementary two person interactions, come to the laboratory to trade in impersonal experimental markets, what do they do? They strive to maximize their own gain, and in this process maximize the joint benefits of the group but without intending to. However, these markets are supported by externally enforced property right rules that prevent individuals from taking without giving in return, and it's the community support for these property rights that enable trade and specialization to occur.

In established commodity and service markets producers incur recurrent, relatively predictable costs, and consumers experience corresponding recurrent flows of value from consumption. But costs and values are inherently private and all such information is dispersed, decentralized among individuals. Command and control economies have generally tended to fail because such information cannot be given to any one mind. But how do we know that the price discovery process in commodity markets yields efficient surplus-maximizing outcomes? Well, we have discovered in controlled laboratory experiments that these recurrent flow markets are incredibly efficient, and these findings have been replicated many hundreds of times by different researchers and laboratories, first all over the United States and then elsewhere in the world. Moreover, the subjects in these experiments are not aware of the group welfare maximizing ends that their actions produce. Each, in pursuing his personal gain, achieves group maximizing benefits that are not part of his deliberate intention.

So what have we learned about markets? What is the unseen function that they serve? First, commodities and service markets are the foundation of existing wealth creation. Each of us earns our income from no more than one or two sources. Yet think of all the hundreds of items that we use or consume throughout the day that are produced by others whom we do not know.

The hallmark of commodity and service markets is diversity: diversity in tastes, human skill and knowledge, natural resources, soil and climate, which in turn account for differences in values and costs that we use to define and motivate gains from exchange in the laboratory.

The power of diversity to be extended and to serve human betterment depends vitally on exchange, both personal exchange in our intimate groupings and impersonal exchange through markets.

Initially, diversity was possible and encouraged through sharing and reciprocity norms in the family, thus, in stateless hunter/gatherer societies, the women and children gathered fruits, nuts, tubers and grains; the men hunted; old men advised in the hunt, fashioned tools, weapons and helped in gathering.

At many times and many places in prehistory, exchange was extended to strangers through barter, and ultimately through the use of commodity money and then finally modern monetary systems. Indeed early humans set the stage for a vast expansion of wealth and wellbeing whenever a tribe discovered that it was better to trade with their neighboring tribes than to kill them. If you kill them, they can't produce something and trade it with you tomorrow, nor can you benefit from their unique skills, learning, art, culture and experience. Similarly, if you let them live but steal from them, they are much less willing to produce more for you tomorrow than if you trade with them today.

Diversity requires freedom, because it is freedom that allows each to be as different as he or she is able and desires to become. Markets in turn support tolerance of freedom. Chile was a country that had little political freedom but opened the economy to freer choice, and this eventually spread to political choice and helped to bring democracy.

Diversity without the freedom to exchange implies poverty: no human, however abundantly endowed with a single skill or a single resource, can prosper without trade. Robinson Crusoe owned an island, but he was poor.

We have need of others and of the diversity they bring to the table if we are to rise above bare subsistence. Through markets we depend on others, whom we do not know or recognize or understand. We know not how and in what ways others contribute to our welfare, and we contribute to theirs. Such are the long subtle chains of interdependence through markets connected by prices. The welfare of each of us depends vitally upon the knowledge and skills of others with whom we trade through markets. Diversity is made possible, productive and permissive of wealth creation through markct institutions.

Without markets we would indeed be poor, miserable, brutish and ignorant; if some were less poor, it would be because of conquest, theft, taking without giving in return, which can be sustained only for as long as there are others to conquer. Markets require consensual enforcement of the rules of social and economic exchange. No one said it better than David Hume over 250 years ago, when he said that there are just three laws of human nature: the right of possession, its transference by consent, and the performance of promises. These are the ultimate foundations of order, with or without formal law, that make possible markets and prosperity.

Notice that Hume's Laws of nature are derived from the ancient Judeo commandments: Thou shalt not steal; thou shalt not covet the possessions of thy neighbor; thou shalt not bear false witness. But these same commandments emerged in other religions the world over as they became norms for sustainable, stable societies. By sustainability, I mean the ability of a community to feed clothe and house itself without transfers from others.

The game of steal consumes wealth without encouraging its reproduction, while the game of trade sustains and grows abundance.

Coveting the possessions of others invites an involuntary state-enforced redistribution of the gains from specialization and trade, endangering incentives to produce tomorrow's harvest perhaps as surely as its theft.

To bear false witness is to undermine community, management credibility, investor trust and confidence, long-term profitability and the personal social exchanges that are most humanizing.

I want to turn next to the topic of stock markets. These markets are inherently far more uncertain than markets for commodities and services because stock markets must anticipate innovations, the new commodities and services of the future. At the time of new innovations, the extent of their subsequent economic success is inherently unpredictable.

In laboratory stock markets, even where fundamental value is well defined inexperienced subjects produce great price bubbles and crashes; if and when they reach a fundamental value, rational expectations, equilibrium, it is through experience. Consequently, the behavior of laboratory stock markets is much more erratic than the recurrent flow markets for goods that we study in the laboratory.

If changing knowledge and technologies are to yield new commodities and services, they require capital. Capital markets allow the users and suppliers of capital to be distinct and more specialized; the savers do not also have to be the entrepreneurs that can grow new wealth from capital investment, and both can gain by exchanging investment for a share of the return, each also bearing the risk of loss.

Stock market bubbles and crashes, such as the one that we experienced just a few years ago, at the end of the decade of the '90s, are not new. Why is this? Essentially, great stock market booms are fueled by new technologies.

For example, in the 19th century the steam engine allowed the steam ship to replace the square-rigger sailing vessel, the railroad to replace the mule team and the stagecoach. Railroad expansion in 19th century America outran the shipping needs of inter-regional trade.

Profitability turned to losses, bankruptcies and consolidations. But out of this 19th century expansion, long-term value was created and retained for the entire economy.

Then at the turn of the 20th century many new technologies emerged. Telephone, electricity, petroleum and automobiles sustained a wave of investment and development. There was over-expansion in response to high profitability followed by declining margins, losses, bankruptcy, consolidation, but long-run value was created and not lost to the economy. Bankruptcy allows the assets of failed managers, human and physical, to be reallocated to successful managers.

If you go back 100 years ago today and look at the early automobile industry which developed in the United States, there were literally hundreds of small manufacturing operations for automobiles.

A third of those small companies were experimenting with battery-powered automobiles. This is 100 years ago. Thomas Edison had invented the battery and he had invented a battery-operated automobile and other people were doing the same thing. All those experiments failed. They failed, for one thing, because the range of these automobiles was not very great, and also there was an inadequate availability of facilities for recharging them, and of course the batteries were very heavy.

Many other of these companies, in fact, most of the other companies at the beginning were based upon the steam engine. The steam engine was a natural thing for people to think of for an automobile because the steam engine had been so successful in the steam ship and locomotive. But there was one major problem: they took a long time to heat up and the creation of an automobile based upon that technology simply was unable to satisfy consumer demand and preferences.

The winner was a long shot: it was the internal combustion engine, but very few at that time anticipated that. Henry Ford was successful in creating an internal combustion engine and in producing a standardized automobile, the Model T. At the time of the First World War, Henry Ford had produced nearly half of all of the automobiles that had been produced and sold.

So my point is, out of this very large number of experiments, entrepreneurs risking their own capital, there were very, very few winners. And of course, the history of the automobile industry since then is marked by constant innovation and change, improvements in the internal combustion engine and all aspects of the motive system.

During the 1970s there was a movement in the United States to protect the automobile industry from Japanese imports. Very fortunately for America, that movement failed. The importation of quality automobiles from Japan helped to motivate and to force American automobile manufacturers to produce a better product, and the American consumer very much benefited from that process. And today we have the prospect that probably within a few years Toyota in Japan will exceed General Motors in the production of automobiles.

I grew up in Wichita, Kansas, and Wichita had 15 airplane manufacturers in 1929, most of them you've never heard of: Lark, Laird, Swift, Knoll, Travel Air. But there were two new company names in 1927: Cessna and Stearman. A decade later Stearman had become part of Boeing and the general aviation survivors were just two: Cessna and Beech. Clyde Cessna had been a farmer who tinkered with farm machinery and was a mechanical genius. Walter Beech got fascinated by the aircraft industry and became a test pilot. These two, Cessna, Beech, and Stearman through Boeing, made Wichita the national center of this new industry and then eventually became an international center for the manufacture of light planes.

The ball point pen today is an almost invisible but classic example of innovation and change. I remember about 1950, I think it was around 1948 or '49, when the first ball point pen came out. They initially sold for $10 back in those days; it was an enormously profitable new product, there was a big rush of entry, falling prices, losses, consolidations, many firms were squeezed out and went bankrupt, but the pen stays, yielding continuing long-term value of which we are not aware, except that we are all a little bit richer as a result. Today in the United States when we buy one of the more popular pens, a BIC pen, it costs 50 or 60 cents, we are unaware that that pen today is far superior to those $10 upstarts over a half century ago.

So picking winners and losers is inherently risky. More than 60% of manufacturing firms have left the industry in their first five years, this is even after they get fairly well established; 80% in their first ten years. Now the 1990s brought an unprecedented volume of new public offerings, and I'm sure that the history of that decade will record an unprecedented failure rate, but also, and much less visible, an unprecedented increase in long-term economic value for the economy. The recent bubble and crash was fueled by new electronic communication, computer, biological and pharmacological technologies. The diesel truck engine is an example of long-term value in the old economy created by companies, some of which are now stressed if not in bankruptcy. Each cylinder is computer controlled for minimizing fuel consumption and meeting tough new emissions standards under all operating conditions.

It is painful for those who risk investment in new technologies and lose, but the benefits captured by other industries, and by the learning and consolidations that leave value for the few winners, are retained as new wealth for the economy that benefits everyone. This is the substance of growth, betterment and the ultimate reduction of poverty. This is why almost everybody is wealthier than were their grandparents.

How can the individual pain be eliminated and the long-term value achieved with a policy fix that avoids the risk of doing more harm than good? We don't know. If you limit people's decisions to make risky investments in an attempt to keep them from harming themselves, how much will that reduce our capacity to achieve technological advancement? The hope of great gain by individuals fuels thousands of experiments in an environment of great uncertainty as to which experiment or experiments will be successful. The failure of the many is part of the cost of sorting out the few that will succeed.

I want next to turn to globalization.

The first long-distance trade between Europe and the Near East allowed us to escape the static dead-end and poverty-ridden Middle Ages. This led to new explorations by stock companies and nation states.

This exploration was driven then by a new technology, and at that time it was the great square-rigger sailing ships.

As commerce spanned the old and new worlds, there was a worldwide exchange of plant and fruit products. The Italian gourmet had not a single tomato until the plant was imported from the new world; nor did the Irish have the potato until one of the thousands of varieties that had grown wild was imported from South America. The diversity of nature was the basis of much wealth creation through exchange. Instead of cutting edge research and development, we had exploration, transportation and transplantation.

So in the 19th century the seas were spanned by steamships, the Continents by the steam locomotive railroads.

Whole regions now began to specialize in different industrial and agricultural products depending upon their natural comparative advantages. The diversified subsistence farm reformed into the cattle ranch, the wheat, barley, corn and rice farms, the milk farm and the chicken farm.

And then, as I've indicated, the latest great thrust in globalization is driven by innovations in computing power, communications and higher-speed transportation. All three have served Internet exchange.

The retail store was once the place where buyers met producers through the intermediary of the merchant who risked the purchase of inventories of what he hoped people would want to buy, and the buyer risked the quality of the goods produced.

This very high-cost way of matching consumers with producers has been challenged by the Internet, where buyers and sellers are matched at practically near zero cost, and new institutions are being created for direct shipment, and for the quality assurance through competition in reputation formation, warranties, liberal return policies. This new dream world of potential profit led to over-expansion as investors threw investment funds at all the retail dotcoms, just as their ancestors a century earlier had thrown investment funds at the railroads.

Current globalization is bringing a new discipline to national governments. Budget and monetary excesses by national governments discourage foreign investment, while encouraging domestic nationals to seek foreign, more stable, investment opportunities.

I visited Mexico two years ago and I learned that monetary policy was aiming at a 3% rate of inflation, but they got 5.5% instead of 3. This is in effect a tax on domestic capital investment, and Mexican investors are motivated to take their capital to better-managed foreign countries.

South American countries can better serve their people by asking how they can learn from Chile to bite the bullet, stabilize currencies, control government spending, privatize government-owned industries, and reduce barriers to free trade.

In particular, there is the need, I think, emphasized by Hernando de Soto for institutional change enabling the owners of real estate and other assets to hold clear fungible titles. Only in this way can exchange value reflect use value and facilitate internal development.

In the United States, there is a very large and liquid market for real estate mortgages, that enables many entrepreneurs to use their homes and mortgages to finance some of their business activities. And when you go to many of the less developed countries of the world, you find that even though there is a lot of real estate investment and development, there is not any good market system for enabling those assets to generate equity investment capital for new businesses.

Not only capital but also people move to where there is opportunity, and this is the essence of creating new wealth and prosperity. In 1978 I first visited New Zealand. I flew into Wellington and I was picked up by a taxi cab at the airport. The driver was very friendly. You can learn a great deal about a country by talking to its taxi cab drivers. I asked him to tell me about his country. Oh, he says, it's a wonderful place. He says, I like living here. He says, of course we have extremely high taxes and I don't like that. He says, as a cab driver I pay half of my income in the form of taxes, but we get all of these free services, free medical and health, free prescriptions, free education, all the way from elementary school to as high as you want to go in universities. He said: my son is going to become a medical doctor; he has finished his medical degree; he has served his residency, and he's ready to start practice. I said, well, that's a wonderful story, I really am happy to hear that. Is he going to practice in Wellington? Oh no, you can't make any money here, he's going to Australia!

What's interesting about that story, it was in 1978, within the next three years New Zealand had a foreign exchange crisis, the country was bankrupt, and that created the movement in the 1980s that led to privatization in New Zealand of a lot of the government-owned industries. And for a while New Zealand was able to reverse this brain drain that it was suffering at that time. Now, what I find particularly interesting about the brain drain is the fact that it is now being reversed in many places in the world. In India, in China, in Ireland, young people, who earlier left to seek opportunities and education elsewhere, many of those people are returning because their home countries are starting to provide better opportunities for those young people.

In conclusion, let me say that commodity and service markets are the foundation of existing wealth creation. The fact that stock markets serve by supplying capital for new consumer products explains why they are inherently uncertain, unpredictable and volatile, tending to bubble and crash. The problem always with new technology is how to manage it to produce products that satisfy consumer preferences, and the costs incurred can be sustained by the willingness of consumers to buy the product. Stock markets are far more uncertain than markets for commodities and services because they must anticipate innovations, the new commodities and services of the future. Globalization is not new. It's a modern word describing an ancient human movement, a word for humankind's search for betterment, and the worldwide expansion of resource specialization, and specialization is determined by the extent of market development.

I believe globalization is a good word, a peaceful word. In the wise pronouncement of the great French economist Bastiat: If goods don't cross borders, soldiers will.

QUESTION 1: People want to grow their savings, and companies want to reduce their debt. The government tried to reduce the budget deficit, and Prime Minister Hashimoto raised taxes, which led to Japan's Great Stagnation in 1997. Each individual did their best to achieve their own profit, but the overall economy failed. The overall economy did not gain the best result. What do you think about Japan's economic situation?

SMITH: Well, of course I don't like to get involved in the politics of foreign countries, but I do believe that there is a movement, as I understand it, toward privatization of some of your industries, as I understand, the post, which is also a very large savings bank, and that's being debated in your country.

I don't follow the macroeconomic policies in Japan. I have a problem being entirely happy with the policies followed by my own government in the United States. We have a very large deficit, a growth in deficit in the last two or three years which I think is going to impede our own growth. The tax cut enabled us to avoid having a general recession following the stock market crash, but I think a very worrisome policy in the United States is the current account deficit.

You know, Japan has been in the past a model of economic growth and development for the world, and I think it's important for them to find the kinds of public policies that can enable Japan to return to that model level of growth and development. Anyway, I wish you best of luck with your political parties.

QUESTION 2: Experiments in the world of physics are trying to test what is happening through tests and trying to verify research and theory, and if the test results are not favorable you would propose new theories. But in economics experiments, do you also have similar experience that through the experiments you've found that certain results would not be indeed as well suited to a certain theory so you would have to change the theory or the assumption? Did you also find similar experiences in your own experimental economics world as well?

SMITH: Very similar, and I want to talk about two examples. The first example is the discoveries that I began. The first experiments I did, beginning in January 1956, were concerned with the performance of markets in which information as to willingness to pay and willingness to accept on the part of buyers and sellers was dispersed among all of the participants, and also private, each individual knew only his own circumstances. It turned out that that was sufficient to give you convergence to competitive efficient equilibrium.

Now that sounds like a great victory for equilibrium theory in economics, except that that theory required complete information on the part of the participants. In other words, the equilibrium theory, which was able to predict what individuals achieved in the market, failed to establish or articulate the conditions, the information conditions that would enable it to come about.

What was missing was an adequate theory of the dynamic process whereby individuals go from their dispersed private information, the exchange of information through a bid-ask process in the market and then converge to a competitive equilibrium. We observed that, but we did not have a good dynamic theory of it, and we still don't. Fifty years later I'm still waiting for the theorists to come up with a better way of explaining that result. It truly was remarkable though that the conditions of supply and demand, which no one knew about in the market, and you have subjects in the experiment that know no economics, they have no sophistication, they do not have complete information, and yet through repeat interaction over time, each person learning to correct his early mistakes, converges to an equilibrium that none of the people in the experiment knows about.

Now that was a rather remarkable thing. And of course my first experience was that economists didn't believe it. And then people began to get interested in doing experiments, and they found that what I had discovered was a true representation of how those kinds of repeat interaction markets work.

So there was a case in which the static theory did better than it had a right to do. It didn't deserve to do that well because the theory didn't really account for the process that we observed among individuals in the experiment.

The other example was what we've discovered in connection with two-person interactions where people are anonymously paired; you don't know who you're paired with, and you and this anonymous counterpart are going to participate, say, let me give you an example of a simple two-move, two-stage game. We call this a trust game, not when we describe it to the subjects, but when we describe it in an article we call it a trust game. Suppose I move first and I'm matched with you. All right, I can do one of two things: I can choose to stop the interaction and I get \$10 and you get \$10, I can choose to opt out, to defect so to speak, and we each get \$10; or I can pass to you. If I pass to you, the \$20 becomes \$40, and you have two alternatives: you can give me \$15 and take \$25, or you can take all the money, the \$40.

Game theory predicts that I shouldn't pass to you because I can see that if I do it's in your interest to take all the money and I'll be left with nothing. Therefore, the equilibrium of that game is for me to opt out and we each get $10. Or if you do this experiment, we've done it with undergraduates, we've done it with teachers, we've done it with chiefs of staff in the U.S. congress and senate, democrats and republicans, and half of the people in the first mover position do not opt out, they pass to the other person. Seventy-five percent of the second movers give $15 to the first mover and take $25; they don't take all the money. Now this is clearly a violation of the kind of standard game theoretic economic way of modeling.

What's wrong? Well, you see game theory assumes that the participants will each act in a very narrow sense of the self interest. Given a choice between two piles of money they will always take the larger pile regardless of the circumstances. Now think about the circumstances in this simple experiment. If I pass to you, that means we both can be made better off. I have incurred the risk that you will take all the money. I do this obviously in the hope that you will give me $15 and take $25 for yourself. Most people, in fact typically about three-quarters of the second movers, appreciate that they should reciprocate my offer to cooperate.

Now most of the subjects that come into these experiments come from a world in which they are traditionally exchanging favors with their friends and associates. There is a phrase that is universal across most of the languages of the world and the phrase is, "I owe you one." Whenever somebody does something for you without even thinking, often people will say, "Thank you, I owe you one." So people all over the world are social exchangers engaged in reciprocal acts of goodness. When an offer to cooperate is denied, we often then use not positive reciprocity but negative reciprocity, we punish that act. And so it's very common for people to engage in punishment acts when people don't return the favor or they decide choose other people for their friends rather than associate with those people.

So my point here is that this laboratory research shows that people in these experiments have a more sophisticated notion of the self interest. It's not that they're not self-interested, but they recognize in simple two-person interactions that each can be made better off by reciprocating acts of goodness. Now of course game theory does predict that when you repeat these games, then you will get this reciprocity and people tending to cooperate. What game theory was unable to anticipate is that the norms of reciprocity are so strong in many societies that people will do it in a single play of the game.

So in experimental economics we have cases in which the theory is not falsified by the results, but even in the case of markets where the equilibrium theory is not falsified, the theory is simply not adequate to explain the dynamics, and in fact economics is still, I think today, very weak in terms of having good ways of describing the process whereby people begin from a very low state of information, exchange information and start to produce outcomes that are not only better but after a while they're optimal. And you can show that they're optimal. And you can also show that no one in the experiment is aware of that. So all of these things can be demonstrated.

So I think there's many differences between economics and physics, but so far as the experimental method is concerned I think there is great similarity.

Also in physics, whenever there is an outcome that is disconfirming of a theory, people ask two questions: is the theory wrong or is there something wrong with the experiment? The same thing is true in economics, and in fact, much of the interaction at professional conferences in experimental economics involves questions of whether it's the theory that's false or whether there's something wrong with the experiment. And of course there's famous experiments in physics where in fact there was something wrong with the experiment.

A very distinguished experimental physicist by the name of Kaufman, in 1905 did an experiment that disconfirmed Einstein's theory of relativity. Ultimately, it was established that there were problems with Dr. Kaufman's experiment, and those problems were solved and the theory ended up being confirmed.

I gave a long answer to your question because the question is at the heart of much of what we do in experimental economics, and of course this university now has a laboratory in experimental economics and is well along in the process of developing the skills and methods that are used to falsify or confirm economic models.

QUESTION 3: In your presentation you talked about commodity markets and stock markets. You said that basically if the market is fully functioning, the wealth of the society will be maximized. I think that was the main point of your talk. What about the labor market? What would you say about the labor market? Here in Japan much labor has been outsourced to China and other East Asian countries, and this has led to unemployment of the Japanese. So as a result of the globalization of the labor market, many Japanese have been deprived of wealth. Do you have any comment about the labor market?

SMITH: Yes. There of course are some features, special features of labor markets that make them different from commodity markets. For one thing, there is much more of a reciprocity relationship between the worker and the employer. It's impossible frequently for the employer to monitor every aspect of the laborers' actions and therefore it's important to have a trust relationship there, and I think good firms do tend to develop that relationship. In fact, I talked about Henry Ford and the automobile industry. Henry Ford innovated what was called the five dollar day at a time when the most wages were around $2.80 a day. And he did that because he was trying to get the best workers to come and work for him and to create a favorable environment for that.

Now outsourcing is politically controversial everywhere, and it was very controversial in our latest presidential campaign between democrats and the republicans. Outsourcing though I think is an excellent example of something that firms have to be allowed to do, and the reason is very simple. Suppose that you prevent Japanese firms from outsourcing where they can get the commodities produced at a lower cost. The competitors of Japanese companies elsewhere in the world can nevertheless outsource, and when they outsource, the Japanese firms, if they're prevented from doing it, run the risk that they will not be able to compete in world markets, they may eventually go bankrupt, and then you lose both the firms, the companies and the jobs.

Politically, there's always a constituency tending to protect yesterday's jobs; there's no political constituency for tomorrow's jobs. When firms outsource and save money it enables them to do two things: one is to lower prices and compete more effectively; the other is to invest in new technologies and better ways of producing new products. And you cannot get the new unless you allow the savings from the old.

You know, in Great Britain when the first textile machinery was invented it was thought that the labor market was going to be very severely impacted, as indeed it was. The Luddites wanted to destroy all the machinery. What's interesting though about the innovations in the textile industry is that that industry was able to expand, produce at much lower cost, and actually make possible an expansion of employment generally in the economies where those innovations were allowed to go forward. In fact, if you look at the history of innovations, there's always new jobs being produced and the employment levels have increased.

The problem I think in economic policy is to help workers adapt and acquire new skills and go into producing new products, and to do that so as to not prevent firms from outsourcing and obtaining the savings which not only that firm benefits from, but all of their customers and everybody they deal with. You know, in the United States, when I was finishing graduate school in Massachusetts, this was 1955, I had been there for three years, New England was still an important center for textile manufacturers. All of that textile industry that started in New York and New England was moving out of the region and into the South. It moved into the South because wages were lower. That movement helped to bring about a replacement of those industries in New England and New York by the higher tech, new innovations that were coming in. The lower wages in the South gradually were bid up. Today the wages are no longer much lower in the south, the textile industry is no longer in the South; it has moved to China and Taiwan and India and places like that. It has moved overseas, and that's exactly the reason why I started with a quotation from David Hume about manufacturers continually moving to where there is lower cost. And of course when they move to lower cost, they tend to raise those wages, and all of that is part of the process of human betterment, increasing output, reducing poverty, and the problem is to find ways to let that happen but allow the adaptation and adjustment to take place in the labor market.

That's what I think your policy should be focused on; how to help people in those industries move into new industries, but don't stop it from happening because that will make you still worse off.

Keynote Speech: "Foundations of Experimental Economics, Economic Design and Applications"

Vernon L. Smith

George Mason University

I want to begin with a couple of quotations. And in fact, here and there in the paper I will be using quotations from David Hume, Adam Smith or Hayek. This is not because I started as a classical economics scholar. It's also not because I started with an interest in Friedrich Hayek, an Austrian economist. The special value of their contributions is what I discovered after having a long career in experimental economics. And in fact, without my experience as an experimentalist I don't think there's any way that I could have been able to appreciate the full significance of these quotations that I'm going to use. I now see them as enormously insightful in terms of what we have learned from experimental economics. What astonishes me is that Hayek and some of these 18th century scholars could have gotten to this level of understanding without doing experiments. I could not have done that.

The first quotation is one of my favorites from David Hume:

> When we leave our closet, and engage in the common affairs of life, reason's conclusions seem to vanish, like the phantoms of the night on the appearance of the morning; and 'tis difficult for us to retain even that conviction, which we had attained with difficulty.

Now current research in neuroscience tells you how brilliant that quotation is because research in neuroscience tells us that we are of two minds, or what I prefer is to distinguish between the mind and the brain, the mind being the conscious self-aware thinking and reasoning we do which is a very, very small part of the brain's activities; the brain is typically on automatic control and this is what enables us to do most of the things we do without having to call upon the intentional and analytical resources of the mind. This distinction comes up over and over again in experimental economics, and we have become more aware of that dichotomy I think as the tools of neuroscience are used to better understand decision making of the kind experimentalists have been involved in over the last 40 or 50 years.

The next quotation is from Hayek, and anyone, which is probably most of the people in this room, who has conducted an experimental economics market, double auction or sealed bid auction, will be able to appreciate the truth and power in this wonderful quotation:

> Rules alone can unite an extended order...Neither all ends pursued, nor all means used, are known or need be known to anybody, in order for them to be taken account of within a spontaneous order. Such an order forms of itself···

You see a dramatic illustration of this principle every time you run a market experiment in the laboratory.

The organizing principle in this address is the simultaneous existence of two rational orders. I shall try to make the case that both orders are distinguishing characteristics of what we are as social creatures; that both are essential to understanding and unifying a large body of experience from socio-economic life, the experimental laboratory, and in charting relevant new directions for economic theory as well as experimental-empirical research.

I want to say also that because the sub-title of this conference dealt with applications to the real world, I have changed somewhat the original talk. I was going to give in order to bring in some of these applications. Accordingly, I will not have much to say about behavioral economics but more about some of the applications of experimental economics, particularly economic design applications, at least in so far as there is time for me to do it towards the end of my talk.

The first concept of a rational order derives from the Standard Social Science Model going back to the 17th century. Hayek called it constructivist rationality, which stems particularly from Descartes, who believed and argued that all worthwhile social institutions were and should be created by conscious deductive processes of human reason. In the 19th century Bentham and John Stuart Mill were among the leading constructivists. Bentham and the utilitarians sought to reform British law and institutions on these rational principles. Mill introduced the much-abused constructivist concept, but not the name, of "natural monopoly." To Mill it was transparently wasteful and duplicative to have two or more mail carriers operating on the same route. He is the intellectual father of the U.S. and other postal monopolies around the world, their resistance to innovation, and their demise in the face of the privatization movement in some countries and the growth of superior substitutes in others.

I noticed in the Financial Times a couple of days ago that Japan is considering privatizing the post and of course it's very controversial; such proposals are always controversial. In this case, as I understand from the Financial Times, the Japanese post is a huge bank, 16% of all Japanese savings is held in this bank. I don't think John Stuart Mill anticipated that in a postal monopoly where people go to buy stamps or mail letters, that it might provide a convenient time for people to put deposits of small sums into savings accounts. I don't know the extent and origin of the Japanese post, but it is the case that postal monopolies around the world have sometimes gotten into the savings business and typically would begin with small savings. But this is an example of how institutions change and adapt in ways that the original constructive motivations could never anticipate.

Mill also could not imagine that it would be efficient for two cities to be connected by two parallel railroad tracks. I would conjecture that by the time of his death in 1873, men with grade school educations had already been launched on the road to riches constructing the first parallel route railroads. Emergent contradictions to constructivist natural monopoly are examples of what I shall call ecological rationality, as detailed below. But much of economic policy, planning, regulation, antitrust and economic theory is still in the grip of Mill's inherently static analysis of natural monopoly.

As Hayek put it, the constructivist view of rationality:

> ... gives us a sense of unlimited power to realize our wishes ... (and) ... holds that human institutions will serve human purposes only if they have been deliberately designed for these purposes... (an institution's existence) ... is evidence of its having been created for a purpose, and always that we should so re-design society and its institutions that all our actions will be guided by known purposes... This view is rooted ... in a ... propensity of primitive thought to interpret all regularity ... as the result of the design of a thinking mind.

In economics, rational predictive models of decision have motivated research hypotheses that experimentalists have been testing in the laboratory since mid-20th century. In market experiments, where cooperation can occur through the coordination function of prices produced by, but simultaneously resulting from, interaction with individual choice behavior, the results are commonly in accord with standard equilibrium models that maximize group welfare. This professional victory is hollowed, however, by the failure of standard theory to predict what turned out to be the surprisingly weak conditions under which the results obtain. Standard theory has also had only limited success in modeling or explaining the emergence of the institutions we copied from the world of practice into the laboratory.

Many economists are either baffled that equilibrium theory works in these private information environments, or continue to think, speak and write as if nothing had changed, that equilibrium theory requires complete or perfect information. Consequently, these experimental discoveries have yet to galvanize many programs of theory modification as applied to these observed market equilibration processes.

Thus, for tractability, Cartesian rationalism provisionally assumes or requires agents to possess complete payoff and other information, far more than could ever be given to one mind. In economic analysis the resulting exercises are believed to sharpen economic thinking, as if-then parables. Yet these assumptions about the economic environment are unlikely to approximate the level of ignorance that has conditioned our evolved institutions as abstract norms or rules independent of particular parameters, which have survived as part of the world of experience. The temptation is to ignore this reality because it is poorly understood and does not yield to our familiar but inadequate modeling tools, and to proceed in the implicit belief that our parables capture what is most essential about what we observe. Having sharpened our understanding on Cartesian complete information parables we carry these tools into the world for application without all the necessary caveats that reflect the tractability constraints imposed by our bounded professional cognitive capacities as theorists.

Herbert Simon expressed the idea that in studying decision making we have to be aware of the cognitive capacities of human subjects and be sensitive to this fact in both our modeling and our interpretation of observations. But he did not talk about my point here which has to do with the bounded professional cognitive capacity of theorists, and the extent to which those bounds have made it extremely difficult for us to better model the equilibration processes that occur in these markets where people have very, very limited information. It's a remarkable fact that as theorists we cannot model the ability of naïve and imperfectly informed agents to achieve unintended equilibrium outcomes.

In summary, constructivism uses reason to deliberately create rules of individual action, and design human socioeconomic institutions that yield outcomes deemed preferable, given particular circumstances, to those produced by alternative arrangements. Although constructivism is one of the crowning achievements of the human intellect, it is important to remain sensitive to the fact that human institutions and most decision making is not guided only or primarily by constructivism.

Since our theories and thought processes about social systems involve the conscious and deliberate use of reason, it is necessary to constantly remind ourselves that human activity is diffused and dominated by unconscious, autonomic, neuropsychological systems that enable people to function effectively without always calling upon the brain's scarcest resource-attention and reasoning circuitry. This is an important economizing property of how the brain works. If it were otherwise, no one could get through the day under the burden of the self-conscious monitoring and planning of every trivial action in detail. Thus, "If we stopped doing everything for which we do not know the reason, or for which we cannot provide a justification...we would probably soon be dead." That quote is from Hayek, in The Fatal Conceit.

No one can express in thoughts, let alone words, all that he or she knows and does not know but might call upon or need to discover for some purposive action. Imagine the strain on the brain's resources if at the supermarket a shopper were required to explicitly evaluate his preferences for every combination of the tens of thousands of grocery items that are feasible for a given budget. Such mental processes are enormously opportunity-costly and implicitly our brain knows if our conscious, planning, modeling mind does not know that we must avoid incurring opportunity costs that are not worth the benefit. The challenge of any unfamiliar action or problem appears first to trigger a search by the brain to bring to the conscious mind what one knows that is related to the decision context. Context triggers autobiographic experiential memory, which explains why context surfaces as a nontrivial treatment, particularly in small group experiments.

Perhaps most of our individual mental activities and accomplishments are not accessible to our conscious awareness; all of them depend upon mental processes that are inaccessible to our conscious awareness. Similarly people are not aware of a great range of socioeconomic phenomena, such as the productivity of social exchange systems and the external order of markets that underlie the creation of social and economic wealth.

These considerations lead to the second concept of a rational social order: an ecological system designed by no one mind, which emerges out of cultural and biological evolutionary processes: home grown principles of action, norms, traditions and what is often referred to as "morality," which is the term used by Hume for our evolved norms and practices. As stated by Hayek, this concept of ecological, as I call it, rationality:

> leads to the insight that there are limitations to what we can deliberately bring about, and...that that orderliness of society which greatly increased the effectiveness of individual action was...largely due to a process...in which practices...were preserved because they enabled the group in which they had arisen to prevail over others.

Those of you who heard Dan Friedman's talk on evolutionary dynamics[1], or who are familiar with any of that literature, know that this is precisely the kind of general hypothesis that underlies some of that work.

Constructivism, however, can use reason in the form of rational reconstruction to examine the behavior of individuals based on their experience and folk knowledge, who are naïve in their ability to apply constructivist tools to the decisions they make; to understand the emergent order in human cultures; to discover the possible intelligence embodied in the rules, norms and institutions of our cultural and biological heritage that are created from human interactions but not by deliberate human design. People follow rules without being able to articulate them, but they may nevertheless be discoverable. This is the intellectual heritage of the Scottish philosophers and Hayek, who described and interpreted the social and economic order they observed and its ability to achieve desirable outcomes. The experimental laboratory provides a tool for testing hypotheses derived from various models of these kinds of emergent orders.

David Hume, the 18th century precursor of Herbert Simon, was concerned with the limits of reason, the bounds on human understanding, and with scaling back the exaggerated claims and pretensions of Cartesian constructivism. Both Hume and Adam Smith argued that the order in life and society follows

---

[1] Chapter 7 of this volume. pp.101 - 118.

from emergent norms and learning born of experience far more than the constructivist designs of reason. To Hume rationality was phenomena that reason discovers in human institutions and practices. Thus, from David Hume, "the rules of morality...are not conclusions of (our) reason."

Adam Smith developed the idea of emergent order for economics. Truth is discovered in the form of the intelligence embodied in rules and traditions that have formed, inscrutably, out of the ancient history of human social interactions. This is the antithesis of the anthropocentric belief that if an observed social mechanism is functional, somebody, somewhere, somehow in the unrecorded past surely must have used reason consciously to create it to serve its perceived intended purposes. It isn't long ago that people believed that some genius in the past invented spoken language. This is the default folk belief in the historic origins of any legacy that is functional. But in cultural and biological evolution, order arises from mechanisms for generating variation to which is applied mechanisms for selection. Reason is good at providing variation but very poor at selection, that is, constructivism is a powerful engine for generating variation, but it is far too limited and inflexible in its ability to comprehend and apply all the relevant facts in order to serve the process of selection, which is better left to ecological trial and error processes.

These themes I want to illustrate and discuss in several examples and I'll be drawing on learning from experiments as well as field observations to illustrate how the contrast between constructive and ecological rationality informs learning from observation. And then toward the end I also want to discuss some of the examples that have been growing out of the field of economic design.

But first, how are the two concepts of a rational order related? Constructivism takes as given the social structures generated by emergent institutions that we observe in the world and proceeds to model them formally. An example would be the Dutch auction or its alleged isomorphic equivalent, the sealed bid auction. Constructivist models need not ask why or how an auction institution arose, or what were the ecological conditions that created it or why there are so many distinct auction institutions. In some cases it is the other way around. Thus revenue equivalence theorems in auction theory show that the standard auctions generate identical expected outcomes, which, if taken literally, leaves no modeled economic reason for choosing between them. But society chooses between them in particular applications. It is important to ask how and why, and to avoid dismissing such learning as perhaps irrational or uninformed.

More generally, using rational principles, as theorists we represent and observe socioeconomic institutions with an abstract interactive game tree. Contrarily, the ecological concept of rationality asks certain prior questions: from whence came the structure captured by the tree? Why this social practice, from which we can abstract a particular game and not another? Were there other practices and associated game trees that lacked fitness properties and were successfully invaded by what we observe? There is a sense in which all ecological systems, whether cultural or biological, necessarily must be, if not already, in the process of becoming rational because they serve in some sense the fitness needs of those who unintentionally created them through their interactions.

Constructivist mental models are based on assumptions about behavior, structure and the value-knowledge environment. These assumptions might be correct, incorrect or irrelevant, and the models may or may not lead to rational action in the sense of serving well the needs of those to whom the models are supposed to apply. As theorists the professional charge for which we are paid is to formulate and prove theorems. A theorem is a mapping from assumptions into testable or observable implications. The demands of tractability loom large in this exercise, and to deduce anything much in the way of results it is necessary to consider both the assumptions and the implications as variables. Few game theorists, building on the assumption that agents always choose dominant strategies, believed this to characterize the behavior of all agents in all situations.

To understand what is, the surviving tip of the can-do knowledge iceberg, requires understanding of a great deal more that is not. This is because of the rich variety of alternatives that society may have tried but that have failed. Nor is there any assurance that arrangements fit for one economic and social environment may be fit for another. In the laboratory we can not only rationally reconstruct counterfactuals, as in economic history, but also use experiments to test and examine their properties. And I think always, as many of you have heard me say many times before, whenever we test outcomes in the laboratory, and the theory fails, we should ask, is it the theory that's wrong or is there something wrong with our experiment? And also, if the theory after retesting appears to be firm, we should then always push to more demanding tests of that theory. I believe any of these models can be brought to the edge of failure by putting sufficiently great demands on them. And the purpose of this methodological exercise is to not find fault either with the theory or the subjects in an experiment, but to better understand the edges of validity or invalidity of whatever the system is that we're testing.

Now I want to look at two examples from the world; the second one also involves some experiments. In my discussion of these two examples I'm going to try to illustrate how constructivist and ecological rationality principles are involved.

The first example is the deregulation of airline routes. Airline route deregulation in the United States occurred in the late 70s and early 80s. Airline route deregulation brought an unanticipated reorganization of the network called the hub-and-spoke system. What's interesting about that is that this seems to be an ecologically rational response anticipated by none of the constructivist arguments for deregulation and predicted by no one. In fact, when it emerged people were wondering how that had happened and wondering why. Nor could it have been uncovered, I submit, in 1978 by opinion surveys of airline managers or by marketing surveys of the airline customers.

Unknown to both managers and customers was the subsequently revealed decision preferences of customers who unknowingly favored frequency of daily departure and arrival times, a preference that had to be discovered through market experimentation. Nonstop service between secondary cities was simply not sustainable in a deregulated world of free choice. The only way simultaneously to achieve efficiency, the demand for frequency of service, and profitable load factors among secondary cities was for the flights to connect through hubs.

Now 25 years later this is starting to change, partly because one of the things that's happening is demand is growing between the secondary cities and some of them can now support higher frequency of nonstop service; also you have these new smaller capacity jets that hold around 90 or 100 people and they have lower break even load factors.

I should note there after deregulation there was no shortage of constructive attempts by business people and startup companies to satisfy what demand there was for nonstop service between secondary cities. What they would do is put on an early morning flight and an evening flight, but people who want to leave at ten in the morning don't want to wait for the evening flight, so they go through a hub. And so unless the services could be frequent enough to satisfy the distribution of people throughout the day and the density requirements for break even load factors, that service was not going to emerge. And this, I think, is a beautiful example of the kind of trial and error service structure that people discovered that's extremely hard to uncover by any other device. If you ask people, if they would like non-stop service between their home city and these two secondary cities, of course they would. What they don't know yet and have no way of telling you, is all of the conditional circumstances under which they would actually use or not use nonstop service. And this is the kind of thing that these ecological processes discover by people entering the airline industry, risking their own capital, trying and failing, and other people learning from that experience.

A second and very troubling example that I want to talk about is the circumstances leading to the California energy crisis. As in other regions of the country and the world, deregulation was effected as a planned transition with numerous political and stakeholder compromises. Many of the stakeholders, most notably the local distribution utilities, did not even comprehend what was in their own best interests. In California, liberalization took the form of deregulating wholesale markets and prices while continuing to regulate retail prices at fixed hourly rates over the daily and seasonal cycles in consumption. The utilities believed that it was in their highest priority self-interest to lobby for and to negotiate an increase in these average retail rates to meet the revenue requirements of previous capital investments that were stranded by competition. The word "stranded" was a term invented to describe the belief that they would be unable to recover the costs of those investments under competition. This preoccupation with the past and with irrelevant average revenue/cost thinking by regulators and regulated alike ill-prepared the state for the consequences of having no contingent dynamic mechanisms for prioritizing the end use consumption of power.

As expected by many scholars, traditional volatility in the marginal cost of generated electricity was immediately translated into volatile intra-day wholesale prices. What was not expected by anyone was that a combination of low rainfall that reduced Pacific Northwest hydroelectric output, growth in demand, unseasonably hot weather, generators down on normal maintenance schedules, and so on, caused the temporary normal daily peaking of prices to be greatly accentuated and to be much more lasting than had already occurred earlier several times in the Midwest and South. Well, events of small probability happened at about the expected frequency, and since there are many such events the unexpected is not that unlikely. When supplies are short and demand unresponsive to price increases, this invites gaming of price determination in the spot market and gaming of ill-advised bureaucratic incentives/rules governing access to constrained lines and retail sale credit. These rules allowed suppliers to fake congestion relief, buy at retail because they could get credit for having some retail demand-this is one of the games Enron played-and then sell at the much higher wholesale price. Of course that's a deregulatory right that should have been used to empower every end use customer who desired to reduce consumption in response to high prices and thereby profit from the savings. But the point is, the rule as it was implemented did not allow arbitrage to improve the system. In fact, it made it worse.

Because of the regulatory mandate that all demand must be served at a fixed price, the planning did not allow for the early introduction of demand responsive retail prices and technologies to enable peak consumption to be reduced. Instead of mechanism design we had fixed retail price design to generate average revenue that was supposed to cover average cost-and it failed. The regulatory thought process is as follows: the function of price is to provide revenue, and the function of revenue is to cover cost. So you estimate cost and revenue and set the price accordingly. Both the regulated and the regulators all think that's fair. But it's the antithesis of the market function of price: the market price, and any seller's corresponding volume, tells that seller the unit cost he or she can afford to pay and not lose money. If the price is already the best you can get and it's below out-of-pocket unit cost, you are probably in the wrong business. Regulators and those they regulate failed to understand this normal market principle and to apply it in the new deregulated regime. For neither management nor the regulators was it natural to think in terms of making a profit from selling less power. That's not the way the thought process happens in a regulated industry. Yet that was precisely the route by which the California distributors could have avoided the loss of an estimated $15 billion: every peak kilowatt-hour not sold at the average retail price would have saved up to ten times that amount of energy cost. This continues to be the most significant characteristic of all power systems in and outside California, and market designs that fail to confront this problem head-on continue to invite disaster.

It is because no one knows what will work best that you have to open up retail energy markets to the field experiment called "free entry and exit." In that experiment, if an energy supply company tries a program and it fails, it is the stockholders that suffer the business loss as there is no regulatory mechanism to pass the cost of failed experiments through to the customers; if the program is successful, stockholders will earn the gains as a return on the capital they risked on the investment. It's that discovery process through which our closet conclusions seem to vanish like the phantoms of the night on the appearance of the morning.

Of course in electric power, one of the major traditional problems has been that people think of electrical energy as something that must necessarily be bundled with the wires. And the regulatory apparatus, beginning with about 1912 or '13, when the first state regulation came into the United States, gave local utilities the franchised monopoly right to tie the sale of energy to the rental of the wires, and that legacy is still with us in the United States. In other words, there's not a separation of the energy supply business from the rental of the wires the way there is whenever you rent a car. When you rent a car you just pay the rental car company for the use of the vehicle. You buy your energy wherever you feel like it.

Now, in the time I have left, which is not very much, I do want to talk about the example you're all familiar with in which experimental market results illustrate both constructivist and ecological rationality. The non-cooperative or Cournot-Nash equilibrium theory has conventionally specified the following precondition for achieving an equilibrium: agents require complete or perfect information on the equations defining the equilibrium; also, common knowledge, all must know that all know that all know that they have this information. In this way all agents have common expectations of the equilibrium and their behavior must necessarily produce it.

But as you know, hundreds of experiments over the past 40 years demonstrate that complete information is not necessary for the equilibrium to form out of a self-ordering interaction between agent behavior and the rules of information exchange and contract in a variety of different institutions that we've studied. Thus economic theory got it right in terms of specifying the equilibrium conditions, but failed decisively in specifying what people needed to know.

Now it's interesting that Nash did not ignore the problem of how a group of agents might actually achieve a non-cooperative equilibrium:

> In an unpublished section, [and its probably significant that it was never published], entitled 'Motivation and Interpretation' of his 1950 PhD thesis he proposed two interpretations with the aim of showing 'how equilibrium points and solutions can be connected to observable phenomena' ... the first one was of a positive kind ...a mass action interpretation...based upon an iterative process, in which boundedly rational players observed the strategies played by their opponents

> drawn randomly from a uniformly distributed population of players and gradually learned to adjust their strategies to get higher payoffs. Nash suggested that the learning process would eventually converge to an equilibrium point, if it converged to anything at all, and remarked that in this interpretation it was unnecessary to assume that agents had full knowledge of the game structure or the ability to go through any complex reasoning process.

Of course it's his second example of complete information that economist all know and have recited ad nausea.

The inherent difficulty in equilibrium modeling of, say, the double auction, is revealed in the fact that so few have even attempted it. Bob Wilson characteristically had the courage and competence to log progress, also Dan Friedman. Dan used an unconventional no-congestion assumption to finesse the Nash-Cournot analysis, concluding efficiency and a final competitive clearing price. Wilson used the standard assumptions of what is common knowledge, such as number of buyers and sellers, in a an environment in which each has an inelastic demand or supply for one unit. He took simple environments-preferences linear in payoffs, no risk aversion or wealth effects, valuations jointly distributed, and agent capacity to "compute equilibrium strategies and select one equilibrium in a way that is common knowledge." Notice that this is an abstract as-if-all-agents-were-game-theorists constructivist model of a thought process that no game theorist would or does use when participating in an experimental market. I know because I have run double auction experiments with game theorists. You pass out the values and the costs, and someone asks: how can one make a decision, you haven't given us enough information? They're still in the closet. Okay, you see how hard it is? They are so bound by their closet reasoning that it's hard to shake them away from it. The answer you give them when they say, "you don't give us enough information," is "don't worry, you can handle it" (laugher), and guess what? They do. But they don't have any idea what it is they do. Their brains can handle the process but their mind cannot model or explain why.

Returning now to Wilson and his model, the model itself generates its own problems, such as degeneracy in the endgame when there is only one buyer and seller left who can feasibly trade, a problem that is not a problem for the experimental subjects who do not know this and see imperfectly informed buyers and sellers still attempting to trade and thereby disciplining price. Extra marginal traders, not knowing that they are excluded from the equilibrium provide opportunity cost endgame constraints on price. Agents of course need have no understanding of opportunity cost at all in order for their behavior to be shaped by it. That's why there's no contradictions between some of the findings in behavioral economics-where they interrogate subjects and find out that they're not good at thinking in an opportunity costly way-and their actual behavior in experimental markets. Demonstrably, people can respond to opportunity costs, but they don't have to understand it in order to function and in order for it to be important.

Wilson implicitly recognizes the weakness in traditional game theory. He implicitly recognizes that this endgame problem is a serious one and he says, and this is a great quotation: "The crucial deficiencies, however, are inescapable consequences of the game theoretic formulation." I suspect that we are squarely up against the limitations, perhaps the dead-end ultimate consequences, of Cartesian constructivism. We have not a clue any more than the so-called naive subjects in experiments how it is that our brains so effortlessly solve the equilibrium problem in interacting with other brains through rules of the double auction and other institutions. I think we model not the right world to capture this important experimental finding. I believe that new tools and new thinking are needed to better model dynamic processes.

And I also want to mention before I stop an interesting contribution which is to demonstrate that an important component of the emergent order observed in these market experiments derives from the institution, not merely the presumed rationality of the individuals. Efficiency is of course necessarily a joint product of the rules of the institution and the behavior of the agents. What Shyam Sunder and his co-authors have shown is that in the double auction market for a single commodity, and we don't know yet how much it generalizes, efficiency is high even with zero intelligence agents, robots, each of whom chooses bids and asks completely at random from all those that will not impose a loss on the agent. Thus agents who are not rational constructivist profit maximizers, and use no learning or updating algorithms, achieve most of the possible social gains from trade using this institution.

Here is a question for you. Does this example, to quote from Hayek, illustrate in a small way those "super-individual structures within which individuals found great opportunities...(and that)...could take account of more factual circumstances than individuals could perceive, and in consequence...is in some respects superior to, or 'wiser' than, human reason..."?

I think one of the important items that should be on the agenda of experimentalists in the future-it's already been on the agenda of some of you-is to try to get a better understanding of what it is about the institutions that have survived that makes them better in some sense than other institutions that may have been tried and failed. Now this involves of course studying what is not, and you can never do that easily with field data, but it's just as easy in the laboratory as it is to study what is. And the hard part and the trick is to conjecture what institutions or arrangements might have lost out in creating the particular institutions that we observe. And of course I think part of that study is to be sensitive to the ways in which some of these institutions may already be under modification and change because of all kinds of changes in technology that are particularly prominent in the last 15 or 20 years.

I don't have time to go into some of the economic systems design issues that I and my co-authors have been working on since we relocated to Washington, D.C., but the important thing I think in economic design, when attempting to create new institutions, is to understand the really important function of test bedding. It's very, very unlikely that our models of a situation where a market is to be applied and it's never been done before, will even come close to getting it right when we take a first cut at design. And that's why it's so important to go through the test bedding exercise. Try it before you fly it.

We did that in electric power but then you're not done in the laboratory because when these systems go into the world they require continuous adaptation. We were involved in liberalization of electricity in New Zealand and Australia, and those experiments took the countries a considerable distance in terms of answering the question of how to get started. But believe me, there has been much change since we started, partly because of changes in technology. What started as half-hour spot markets are now being updated every five minutes in Australia. And a huge number of other changes, I couldn't even list them all, there are so many, that come out of practice and experience that feed back into the changing of the rules.

But it's better to make the initial set of mistakes in the laboratory where it doesn't cost much. If you make them out there in the field, they're not only very costly but you get interest groups built up around the bad design and then it's very hard to change.

Question 1: What is the right way to cope with market failure such as public goods provision?

SMITH: This is a very nice question because I can't answer it.

There is no general answer to that as far as I know. The devil is in the details. Deregulation of airlines is a much different process than deregulation of electricity, and all market designs have to be integrated with the engineering and technical realities of whatever the industry is and you have to be very sensitive to those details.

Also, it's very easy in the abstract to imagine public good problems that in fact don't exist. In electric power, all the suppliers are sharing a common wire or set of wires. Losses in the system vary and go up as the square of the total amount of power going through. That looks like there's a huge externality going on that's impacting each individual, but everybody is putting their bids and asks in simultaneously and all of the constraints are satisfied, which is exactly what the algorithms do. Then the prices, nodal prices take into account the effect of each person's bid on the opportunity cost of others and there is no problem.

But you don't arrive at that solution with just pencil and paper easily, and even if you did it needs to be tested. And that's exactly what we did, and I can assure you, I can show you lots of experiments in electric power where people are sharing common resources, and it looks like there's some incredible externality there that would cause problems, but the market is efficient. There's no money being left on the table. So the answer is always about looking at the details.

And also, it's not a simple matter of coping with market failure, such as public good provision. There's also the problem of coping with government failure for public good provision. Why does anyone think the government is getting it right? They have the same lack of information that the private sector has. Information is inherently decentralized in all social systems, and the problem is to find ways of aggregating that information into the production of outcomes that ideally are surplus maximizing, but in any case, you want them to be better than the alternatives that you can think of.

So the failure of social systems is not somehow a special problem for markets, it's a pervasive problem in social organization. And I think the design issues need to be confronted and, given those designs once you get a sense of how those design issues can be approached, then ask what combination of private or public control will best implement whatever the issues are. The answer you look for on the public side is the specification of property (human) rights to take action so that the private side can search and achieve the right answer without any one mind in charge.

QUESTION 2: You spoke of ecological rationality. Does this mean optimality? Is there some presumption or something stronger than a presumption that this ecological evolutionary system will create a surplus maximum?

SMITH: Well I think if money is being left on the table somewhere there's likely to be some attempts to sweep it up and to make the system more efficient. But I wouldn't want to insist that it has to be optimal as traditionally formulated; you also have to deal with the implementational and operating costs costs for the system. But the traditional modeling is a way to start recognizing that the experiments will have a separate life that leads to variations on the rules whose performance can be compared with the baseline tradition.

You know, I think our discovery in the laboratory that these equilibrium outcomes can be obtained remarkably easily through repetition may give us over confidence in our ability to generalize. And that's one of the reasons why I textbfasize the fact that economic theory is very lucky to have gotten so much right, and you don't want to forget the fact that they got the information conditions completely wrong.

What I like about the exploration of evolutionary type models is that you have other criteria besides optimality that can be used in comparing alternative arrangements, although it's interesting that we find in nature economizing principles everywhere. In electric power systems why does power follow Kirchhoff's and Ohm's laws. Well, it minimizes the heat loss. It's the nature-ever-economizes principle. So the electrons that know nothing about constructivist models have no difficulty dividing between two parallel lines so as to make the marginal costs of losses in this line equal to the marginal cost of losses in that line. And that's a principle from nature, not an economic theorist. Lord, J. J. Thompson discovered that principle. He designed the first electric power lines and discovered that nature economizes.

QUESTION 3: Is it necessary to oppose spontaneous order and deliberated order? Do you think that in the future a continuum of institutional forms will emerge for better regulation in cooperation between regulators?

SMITH: Well, the way I think of it is that people are making constructivist proposals all the time. That's what a new entrant in the airline industry does who believes that he can provide nonstop service between the secondary cities and make money. He believes this and he gets other investors who believe it too, and he drives it and he finds that he can't.

Now the point is that what we observe at any one time out there is the state of the system which involves people trying these constructivist schemes and then ecological criteria are saying, no, you can't make money doing that, and so they quit. And of course, Shyam Sunder touched on this in his talk, and it's what motivated my question about the issue I brought up about stock markets, because what Shyam was saying was that capital markets are really about trial and error experimentation with managing new technologies.

You've got a new innovation, a new technology, you've got managers that want to find a way to create a profitable product from the tech nology. And that's exactly what happened in the 19th century in the United States with the steam engine. It's what happened with the automobile, airplane, electric power, telephones and so on. You get innovations, managers, people trying to figure out how they can create profitable new products. There is no experience to go on because by definition, even though it may have some elements of old systems in it, people are trying completely new things, and that system is inherently uncertain. If you could figure out how to solve these kinds of problems without going through the cost involved in the trial and error process, you could easily be rich; hoards of people underestimate the cost of discovery, which is exactly what the last bubble was all about. And I believe though that if you look at these innovations and the booms that have occurred in the past, you typically see long-term value coming out of them in which a fairly small number of breakthrough managerial efforts enable new products to come out, and it may be only 1 or 2% of all the things that have been tried. But that sets the stage for a huge creation of long-term value.

And we don't model those kinds of processes very well. Take the Bayesian updating scheme which enables you to learn efficiently from samples. What does that model do? It begins by assuming that you can exhaust a description of the states of nature. That assumption avoids confronting a problem that most people and most systems don't know how to solve but that's what most of the investment is all about. You're not drawing balls from urns that have a known composition. Life is a world in which you're drawing balls from an urn with red and black balls, and all of a sudden you get a yellow one or a polka-dotted one, and that's not part of the Bayesian apparatus.

I once talked to John Harsanyi about this, and the question I put to him was whether you could always reserve some of the probability mass for surprise states because life and business is about learning from samples what the states are. It's not simply learning about the probabilities that certain known states will happen. And that's the problem that the business person faces when he's deciding how to allocate his investment budget. And you know, there's these constant surprises that cause firms to go bankrupt.

I'm not saying it's impossible to apply some probability calculus to this, but I don't know of any attempt to do it that has solved the kind of problem I talked about. John Harsanyi didn't have a solution.

Part II

The Past and Future of Economic Experiments

Behavioral Decision Making at 50: Achievements, Prospects, and Challenges *

Robin M. Hogarth

ICREA and Universitat Pompeu Fabra

Summary. Many people consider that Ward Edwards' 1954 review paper marks the beginning of behavioral decision making or the study of how people make decisions. Fifty years after Edwards' paper, it is illuminating to reflect on the progress of the field over the last five decades and to ask what the next fifty years might have in store. To do this, I identify ten major ideas or findings that have emerged to date. These are: (1) that judgment can be modeled; (2) bounded rationality; (3) to understand decision making, understanding tasks is more important than understanding people; (4) levels of aspiration or reference points and loss aversion; (5) heuristic rules; (6) adding and the importance of simple models; (7) the search for confirmation; (8) the evasive nature of risk perception; (9) the construction of preference; and (10) the roles of emotions, affect, and intuition. I further identify major challenges currently facing the field. These include linking knowledge to the growing body of work in neuroscience, developing methodologies that can generalize experimental results, having more impact on helping people make decisions, and extending collaboration with other disciplines in the social sciences.

1 Introduction

The area of research called "judgment and decision making" involves researchers from several disciplines and especially psychology, economics, and statistics. It has also permeated many applied fields such as accounting, marketing, and medicine. Central to this research is the interplay between two kinds of models: on the one hand, normative models of how decisions should be made; and, on the other hand, descriptive models of how people actually make decisions. The research is typically pursued because, first, judgment and

* The author also thanks Paul Slovic for constructive comments and acknowledges the support of the Spanish Ministerio de Educación y Ciencia . While revising this paper after the conference, the author learned of the death of Ward Edwards in February 2005. He hopes that this paper can be thought of as a small tribute to Edwards and the work that Edwards inspired.

choice are ubiquitous and important human activities that people find interesting and, second, there is a general expectation that scientists might help people make better decisions.

People (and animals) have, of course, made decisions since the dawn of time. However, the formal study of decision making can probably only be dated to the year 1654 when Fermat and Pascal started to correspond about how to use mathematics in certain forms of gambling [1]. One interesting dimension about this correspondence was that a "good" strategy could be formalized and that this was not necessarily what people were actually doing.

Three centuries later, Ward Edwards [21] published what is generally considered the seminal contribution to the modern study of judgment and decision making. This was an extensive review of studies from economics and psychology in which human behavior on decision making tasks was compared to predictions generated by models from economics and statistics. The timing of this review was propitious in that testable theoretical formulations of decision making models had been and were about to be published (von Neumann & Morgenstern [102]; Savage [84]) and American psychologists - recently freed of the shackles of behaviorism - were about to engage in the so-called cognitive revolution which allowed them to ask questions about how people think (cf. Bruner, Goodnow, & Austin [4]). In addition, the advent of usable computers added two dimensions. One was the ability to exploit computational capacity. The other was the use of the computer as an information processing metaphor of the mind.

In this paper, I reflect on progress made in the past five decades and identify challenges and opportunities facing the field at the present time. To do this - as well as to comment on historical developments - the paper contains two main sections. In the first, I identify ten key ideas or findings[2] that I believe represent the most significant advances. In the second, I outline what I see as the major challenges. In short, I shall conclude that the field is currently in good health and that future generations of researchers face wonderful opportunities.

In terms of the sociology of the field, it is important to note that the leading professional society in the United States recently celebrated its 25th annual meeting with some 350 scientists in attendance. Its counterpart in Europe has been holding meetings every two years since the early 1970s. There is a specialized medical decision making society (that hosts annual meetings) and sessions on judgment and decision making are typically held in meetings organized by social and experimental psychologists, management and organization scientists, economists, and scholars in accounting and marketing. There are

[1] This is, of course, a very Western perspective. I would be curious to know, for example, whether Chinese scientists had not made similar investigations. They were certainly much more advanced than their Western counterparts in the Middle Ages.

[2] The terms "ideas" or "findings" are used in quite a broad sense. Also, this presentation benefits much from previous thoughts on the topic (Hogarth [47]).

two major journals (Organizational Behavior and Human Decision Processes and the Journal of Behavioral Decision Making) but the majority of research - and often the seminal papers - is published in leading journals within disciplines, e.g., of psychology and economics. Researchers who identify with the field of judgment and decision making tend to work at the intersection of fields in departments of psychology or business.

Whereas judgments and decisions ultimately result in observable actions, researchers tend to focus efforts on the distinctive components of inference and preference. Using the economic model of choice as a conceptual metaphor, work on inference covers processes of judgments or beliefs; work on preferences deals with issues of utility, e.g., risk. An important question in the research focuses on deciding when a judgment or choice is or is not "good." Two criteria are used: "correspondence" or the extent to which judgments or decisions are in correspondence with some empirical reality (e.g., the accuracy of forecasts of, say, the weather, or somebody's actions); and "coherence" or the extent to which a judgment or decision is consistent (i.e., coherent) with dictates of normative models (e.g., the conjunction rule in probability theory, Tversky & Kahneman [99]; or invariance of expressed preference under different modes of elicitation, Lichtenstein & Slovic [71]; Grether & Plott [37]). Whether researchers should use correspondence or coherence criteria in evaluating judgments and decisions has been the topic of heated debate (Hammond [40, 41]).

The main topic of this book is experimental economics. Although I have done some work in this area, it is not my intellectual home. I hope, nonetheless, that my comments will be seen as relevant to this important and growing area of social science.

2 Ten Key Ideas or Findings

I now identify my choice of the ten key ideas or findings and ask the reader's indulgence for my personal and obviously subjective selection.

1. That decisions can be modeled. As revealed by Edwards's review, prior to 1954 several studies compared the decisions people made with the outputs of simple expected utility models. The goal of this work, however, was not to model people's choices as such but to assess whether they were consistent with expected utility. The mindset of the time was to make "as if" interpretations of the consistencies and inconsistencies observed.

In 1955, Kenneth Hammond published a paper in the Psychological Review that opened the path to modeling psychological processes of judgment. Hammond's insight was to make the connection between processes of perception (as conceptualized by Brunswik [5]) with those of judgment. Both, according to Hammond, involved inferences concerning a criterion made on the basis of cues. (Consider, for example, the similarity between the tasks of recognizing a friend - a perceptual judgment - and inferring the age of a person you have just met.) Thus, assuming that people's judgments are not inconsistent, it is possible to model the process by expressing judgments as mathematical functions of the characteristics of the informational cues on which they are based. In this case, one does not need a normative model of choice (such as expected utility theory) with which to compare outcomes (i.e., a coherence criterion). Instead, one can compare outcomes with empirical phenomena and investigate why the person does or does not achieve correspondence (i.e., depending on how cues have been weighted, inconsistencies, and so on).

One can, of course, argue that such mathematical descriptions are not really "processes" (I shall return to this issue below) but despite obvious difficulties the modeling approach has proven most useful (cf., Hoffman [44]). First, as exploited by Hammond, Brehmer and others, much has been learned about processes of learning and conflict resolution (Brehmer [3]). Learning, for example, can be quite sensitive to what type of feedback is received (Balzer, Doherty, & O'Connor [1]) and inconsistencies in the expression of judgmental strategies can greatly impede the resolution of conflict (Hammond [41]).

Second is the fact that cognitive models of experts are often better predictors of given criteria than the experts themselves. This finding, known as "bootstrapping" (Goldberg [35]), was important for at least two reasons. One is the notion that a model can extract the essence of a judgmental strategy from a noisy process. The other is the practical implication. Even if no criterion variable is available (imagine predicting success in a new kind of occupation), a model of an expert's judgments is likely to provide more accurate predictions than the expert. In more modern language, bootstrapping models are parsimonious forms of "expert systems" that can be used in prediction tasks even though their inherent simplicity belies the apparent complexity underlying most expert systems.

Models of processes do, however, raise the issue of what is meant by an adequate process model. Clearly, models must meet predictive tests with external criteria. However, most psychologists would argue that they should do more. This raises the issue of the level at which processes are modeled and what ancillary data should and could be marshaled in their support. One clue to this question was provided by Einhorn, Kleinmuntz, and Kleinmuntz [26] who showed that simple linear models are capable of mimicking the more complex processes that are typically represented by so-called process-tracing models. This, in turn, suggests that process-tracing models may be mimicking more complex, underlying psycho-physiological models of information processing. Indeed, recent work in neuroscience attempts to link process models of decision making with levels of activation that are measured in different areas of the brain (see below).

The practice today is to support (or refute) models of decision making by providing measures of process (e.g., reaction times) that are consistent (or inconsistent) with the use of different models. Indeed, the development and availability of computers has vastly facilitated the collection of data and illuminated many process issues (Payne, Bettman, & Johnson [82]). Parenthetically, it is worth noting that prior to roughly 1980 many decision researchers tended to look down on data collection methods that did not use objective statistical tools (such as regression models). However, today most researchers acknowledge the importance of ancillary measurements of process and use multiple methods.

2. Bounded rationality. For many researchers, the concept of bounded rationality introduced by Herbert Simon is perhaps the key concept in the study of judgment and decision making. As stated by Simon,

> The capacity of the human mind for formulating and solving complex problems is very small compared with the size of the problems whose solution is required for objectively rational behavior in the real world - or even for a first approximation to such objective rationality (Simon [90], p. 198).

The importance of Simon's ideas was explicitly recognized by Daniel Kahneman in his Nobel prize lecture:

> ..., we (*Tversky and Kahneman*) explored the psychology of intuitive beliefs and choices and examined their bounded rationality. Herbert A. Simon [88, 92] had proposed much earlier that decision makers should be viewed as boundedly rational, and had offered a model in which utility maximization was replaced by satisficing. Our research attempted to obtain a map of bounded rationality, by exploring the systematic biases that separate the beliefs that people have and the choices they make from the optimal beliefs and choices assumed in rational-agent models. The rational-agent model was our starting point and the main source of our null hypotheses. (Kahneman [58], p. 1449, names in italics added).

It is clear that the concept of bounded rationality has been critical to the study of judgment and decision making in emphasizing the importance of cognitive limitations relative to the unlimited powers assumed by maximizing models in economics. However, it is also the case that the concept has been the subject of misinterpretation.

For example, in an extensive review written for economists and entitled "Why bounded rationality?" Conlisk [14] cites many studies (and reviews of studies) where people's judgments and decisions violate economic reasoning. These studies, according to Conlisk, attest to the importance of bounded rationality. However, just because people are observed to make systematic errors, it is not clear that these are a consequence of bounded rationality per se, i.e., of limitations in cognitive capacity (note the exact wording of the Simon quote above). For example, imagine that when using past performance to predict future performance of an athlete, a person fails to allow for regression toward the mean. Is this due to lack of computational capacity? If it is, then one should predict that no human could make the correct judgment.

And yet, we know that making a cognitive adjustment to a prediction to allow for regression effects requires little cognitive effort. All it really requires is the ability to recognize that one is dealing with a case involving regression effects, something that can either explicitly taught or learned through experience (Nisbett, Fong, Lehman, & Cheng [77]). Moreover, our ability to recognize (situations, faces, patterns, and so on) is one cognitive ability that could almost be considered unbounded (Goldstein & Gigerenzer [36]).

In this sense, I would also argue that Kahneman was selling his and Tversky's work short by describing it as "mapping bounded rationality." True, they were exploring differences between systematic responses and normative prescriptions. But, most of their enlightening experimental demonstrations did not exploit bounds on cognitive processes. Indeed, the simplicity of their paradigmatic problems characterized their appeal and facilitated their diffusion through the scientific community[3]. Indeed, as Kahneman ([58], p. 1469) himself acknowledges, many erroneous responses to these problems arose because people simply answered a different (or easier) problem than that asked by the experimenters. Sure, cognitive processes have a cost but this does not mean that they are "bounded" relative to the task at hand. In many cases, researchers have been studying behavior that could be more aptly labeled as "unboundedly irrational."

Finally, Simon's insight that decision makers do not possess unbounded cognitive capacities was fundamental in allowing people to conceive of alternative models of decision making. The spirit of these models, however, is not just to document the consequences of bounds but a search to make models more veridical from a descriptive viewpoint. Indeed, this viewpoint permeates practice in what is now called "behavioral economics."

3. To understand decision making, understanding the task is more important than understanding the people. In many ways, much of the research in the field attests to this statement. At one level, it is related to the idea of bounded rationality in that organisms do not have, a priori, unbounded computational resources that allow them to deal with an infinite variety of tasks. How then, do they have the capacity to deal with different tasks?

The explanation is evolutionary in nature. It was emphasized by Brunswik [5] and echoed by Simon [89]. The key idea is that the specific response (or decision) mechanisms that organisms develop are conditioned by the kinds of environments they have encountered in the past. Thus, to understand what organisms can and cannot do, it is important to study what they have been required to do by the situations with which they have been confronted. Organisms adapt to the environments they inhabit and thus understanding the

[3] This is not true, of course, of all the Tversky-Kahneman problems. Some, for example, explicitly exploit difficulties of computation to hide the detection of dominance in choice problems (see, e.g., Tversky & Kahneman [101]).

nature of the tasks they face leads to understanding how they have learned to cope with those tasks.

There are many examples of these principles in the literature. Let me cite a few. First, consider the extensive work conducted by John Payne and his colleagues on "contingent decision making" (Payne [80]; Payne, Bettman, & Johnson [82]). These researchers have provided extensive documentation of how simple decision strategies reflect and adapt to environmental characteristics. Second, there is a substantial literature on various forms of preference reversals whereby different presentations of the same judgment or choice problems lead to systematic changes in decisions. I refer here to the classic preference reversal phenomenon that has been studied in economics (Grether & Plott [37]), framing effects in choice (Kahneman & Tversky [61], see also below), evaluating options together with alternatives or in isolation (Hsee, Loewenstein, Blount, & Bazerman [56]), and so on. Third, there is a class of problems in inference where stating problems in terms of frequencies instead of probabilities induces systematic changes in responses (Tversky & Kahneman [100]; Gigerenzer & Hoffrage [32]).

One feature that almost all these problems have in common is that people are sensitive to the information processing costs of different task structures (or problem formulations). They tend to take the "less costly" options. However, and as noted above, this does not mean that the more costly solutions are beyond computational bounds.

4. Levels of aspiration or reference points and loss aversion. A key notion discussed by Simon [88] is that of levels of aspiration or reference points (see also Siegel [87]). In Simon's approach, aspiration levels were seen to be a mechanism by which decision tasks are simplified so that people can use "satisficing" mental strategies (his specific example concerned setting the selling price of a house).

From a broader psychological perspective, the aspiration level or reference point idea is related to the more pervasive perceptual phenomenon of figure and ground. In any given environment, we need to focus attention and tend to do so on what is salient thereby simultaneously separating the perceptual field into what is salient (figure) and what is not (ground). Although this is clearly a generally adaptive mechanism (it is hard to think of an alternative), it comes at a cost. Salience is relative and what is salient can change with shifts in the ground.

The literature is replete with examples of this phenomenon in tasks involving both choice and inference. For the former, consider demonstrations of Weber's rule, e.g., $10 might seem a large sum when paired with $20 but small when paired with $2,000. For the latter, the attribution of cause is greatly affected by assumptions concerning the causal background (Einhorn & Hogarth [25]; McGill [75]).

There is no doubt that Kahneman and Tversky's [61] prospect theory paper has been one of the highlights of the total research program. In my opinion, the key insight in this was linking the notions of reference points and loss aversion, i.e., that "losses loom larger than gains." By itself, loss aversion is a critical concept. However, allowing the definition of gains and losses to vary as a function of reference points or aspiration levels facilitated all kinds of predictions of so-called framing effects going from stunning switches in risk attitudes in laboratory experiments to observations in field studies (Kahneman & Tversky [62]; Johnson & Goldstein [57]).

5. Use of heuristic rules. No discussion of the field of judgment and decision making could avoid reference to so-called "heuristic" rules. The use of the word "heuristic" in this sense originates with Tversky and Kahneman [99] in their famous Science paper on "heuristics and biases" (see also Kahneman, Slovic, & Tversky [60]). The claim made by Tversky and Kahneman was quite simple and measured: In making certain kinds of judgments under uncertainty, people use specific decision rules that, although, often accurate and useful, can sometimes lead to systematic errors relative to normative models.

Tversky and Kahneman specified three heuristics (representativeness, availability, and anchoring - and - adjustment) as well as simple examples of how the use of each heuristic could lead judgment astray. The examples were incredibly compelling and, I believe, it was this fact that turned many scientists into either advocates of heuristics and biases (e.g., Nisbett & Ross [78]) or confirmed skeptics. Among the objections raised against the approach we find two main positions: Tversky and Kahneman used inappropriate normative models in assessing biases (see, e.g., Cohen, [13]) and their heuristic "models" were not precisely specified (Gigerenzer [30].) My own view is that Tversky and Kahneman had developed incredibly important insights about how people assess uncertainty. What was missing was an environmental theory that specified more precisely the interaction between heuristics and tasks (see Hogarth [46]; Kahneman & Frederick [59]).

On a historical note, it should be added that Tversky and Kahneman's early work on biases induced an important change in work on judgments under uncertainty. Before their investigations, most work on this topic centered on how well people's judgments did or did not match statistical reasoning. For example, in 1967, Peterson and Beach had published an influential review on this topic entitled "Man as an intuitive statistician" and Ward Edwards and his colleagues had conducted extensive investigations of people's ability (and inability) to make judgments that were in accord with Bayes' theorem (see, e.g., Edwards [22]). Tversky and Kahneman shocked their readers with their *how well* results (which were bad!) but reoriented investigations by asking *how* people make judgments under uncertainty. The focus became the process that produced the outcomes.

That simple heuristics can also produce good judgments and decisions has been the subject of work by Gerd Gigerenzer and his colleagues (Gigerenzer & Goldstein [31]; Gigerenzer, Todd, & the ABC Group [34]). They have shown empirically how certain simple rules of decision making can perform remarkably well compared to more complex, algorithmic benchmarks. However, as yet their work is incomplete (in the same manner as that of Tversky and

Kahneman) in that they have no theory of how, why and when environmental factors influence the performance of their models (cf., Hogarth & Karelaia [52, 53, 54]).

6. Adding and the importance of simple models. Perhaps one of the most important, practical results is that, compared to unaided human judgment, simple linear models are remarkably effective at making predictions. This has been demonstrated time and again, and in many different contexts (Meehl [76]; Sawyer [85]; Goldberg [35]; Dawes, Faust & Meehl [18]; Kleinmuntz [66]; Grove et al. [38]). Although cases have been reported where humans have outpredicted models (see, e.g., Libby [70]), these are exceptional.

At one level, these findings might be attributed to the notion of bounded rationality in the sense that the human mind is not as efficient as a computer in making calculations. On the other hand, the fact that simple models exhibit superior predictive ability is also a statement about our lack of knowledge concerning the environments in which predictions are made. For many tasks, it turns out that although linear models have higher correlations with the criterion than human judgments, the correlations are not that impressive (see, e.g., Camerer [7]). In addition, people typically want to use more information (i.e., variables) in their judgments than the models.

In addition to the superiority of simple models over human judgment, much work shows that simple linear models are often superior to more complex algorithms for predicting important criteria (Dawes & Corrigan [17]; Einhorn & Hogarth [23]). I return to this below.

The findings of model superiority have not always been embraced with the enthusiasm one might suspect given their practical potential. One reason, I believe, is that investigators have been too quick to celebrate the apparent superiority of models and have failed to understand the conditions under which human judgment can be used as an adjunct to or even instead of model predictions. On the other hand, several researchers have recognized that the issue is not one of "models vs. humans" but of how one can best combine human and statistical models (see, e.g., Blattberg & Hoch [2]; Kleinmuntz [66]; Yaniv & Hogarth [105]). Paradoxically, it could be argued that fairly complex expert systems built within the cognitive science tradition have been better accepted than the simpler models developed in judgment and decision making research. It should be noted, however, that whereas researchers in the latter field have been quick to pinpoint deficiencies in human reasoning, the builders of expert systems implicitly revere the outputs of human judgment by holding it up as a criterion (cf., Camerer & Johnson [8]).

Above, I referred to the work of Gigerenzer and his colleagues. Contrary to the simple models discussed above, their models do not involve averages or sums of several variables but tend to be lexicographic such that decisions frequently depend on a single variable. Thus, the key aspect of their work involves decisions to "choose the best" and "ignore the rest." An important implication of their work is the finding that there are many situations in which models that are "fast and frugal" (i.e., easy to implement and using little information) are quite effective[4].

7. Search for confirmation. In their classic studies of human inference, Bruner, Goodnow, and Austin [4] noted what they called the "thirst for confirming redundancy." This referred to the fact that in solving inferential problems, people have a strong tendency to search for information that confirms their favored hypotheses. This tendency was brilliantly exploited by Wason's classic four-card and "2-4-6" problems that demonstrated that people have a strong proclivity for seeking confirming evidence (Wason [103, 104]).

These findings led to two interesting developments. The first was a purely theoretical investigation by Klayman and Ha [65]. Accepting the fact that people do have a tendency to seek confirmation, they questioned previous research by asking whether, in fact, this was dysfunctional. Specifically, they conducted a task analysis of the effectiveness of a decision strategy that they labeled the "positive test heuristic" - i.e., a strategy that seeks to prove that one's ideas are correct - and showed that, under a wide range of environmental conditions, this is optimal. In effect, they argue that the task used by Wason was a special case and not representative of most inference tasks.

The second development was to question the experimental generality of the Wason task. Specifically, consider the classic Wason task:

> Imagine that you have in front of you four cards that show (from left to right) the letter A, the letter B, the number 4, and the number 7. You are informed that each card has a letter on one side and a number on the other. You are also informed of the rule, "If a card has a vowel on one side, it has an even number on the other side." Given the four cards in front of you, which and only which cards would you need to turn over to verify whether the rule is correct?

As is well known, the most frequent answers to this question are the letter A and the number 4 or just the letter A. In testing a rule of the form if p then q, one should test A (to test that p and q do both co-occur) and 7 (to verify that p and not-q do not co-occur). For most people, however, this problem is quite opaque.

Now consider the following variation of the problem that has exactly the same logical structure:

> Imagine that you work in a bar and have to enforce the rule that, in order to drink alcoholic beverages, patrons must be over twenty-one

[4] Gigerenzer and Goldstein [31] also document intriguing examples of what they call the "less is more" effect, i.e., where the predictive ability of people with little knowledge is better than that of people with more knowledge. Interestingly, earlier Simon [91] noted that in a world where attention is a scarce resource "information may be an expensive luxury, for it may turn our attention from what is important to what is unimportant. We cannot afford to attend to information simply because it is there" (Simon [91], p. 13).

> years old. You observe four "young" people in the bar: the first is drinking beer; the second is drinking Coke; the third is twenty-five years old; and the fourth is sixteen years old. Whom do you check in order to verify that the rule is being enforced? (Gigerenzer & Hug [33]).

Most people have little difficulty answering this question correctly, i.e., they would check the first and fourth persons. What happens here, of course, is that the context of the problem structures the problem for the respondent such that the "correct" solution seems obvious. Exactly why this happens is a controversial issue in the literature. However, I shall return to this example below.

8. The evasive nature of risk perception. What is risk and how do people perceive it? Forty or so years ago, most people thought of risk in strictly consequential terms, i.e., the probability of a negative event occurring and the amount of the associated loss. Thus, when explaining risks of, say, new technologies to the public, the most important issue was to ensure that probabilities of loss were "correctly" understood. It was also thought that studying how people deal with gambles in laboratory settings would illuminate how they conceive and face risks in everyday life.

Starting in the 1970s, these ideas changed dramatically. First, the work on "heuristics and biases" questioned whether people could understand the meaning of probabilities in risk analysis (Slovic, Fischhoff, & Lichtenstein [96]). Second, a large research program, that involved thousands of respondents from different walks of life and countries who answered many batches of questions, revealed new insights into the nature of risk. Specifically, this psychometric approach factor analyzed assessments of many risks and revealed two important dimensions or factors (Slovic [93]). One is the extent to which a risky activity or technology induces a sense of dread, e.g., nerve gas accidents induce a lot of dread; caffeine and aspirin do not. The second factor is the extent to which the technology is unknown or unfamiliar to people, e.g., auto accidents do not represent an unknown risk; however, DNA technology does.

A major impact of these studies is the realization that risk - as understood by people in their everyday lives - is a complex, multidimensional concept that can not be described simply by, for example, postulating different shapes of utility functions. Recent work in this area, therefore, has sought to understand how people come to assess risk in different circumstances, how information about risk can best be transmitted, and the manner in which it is or is not perceived differently by different groups, e.g., substantive experts vs. novices, or men vs. women (see Slovic [94]). The notion that risk is heavily dependent on feelings has gained increasing recognition even though it cannot be said that this has led to elegant models of behavior (Loewenstein, Hsee, Weber, & Welch [73]) - see also below.

9. The construction of preference. Within the normative framework, people make decisions according to preferences that are conceptualized in the form of utility functions. In an as-if approach, it is convenient to imagine that people simply look up the appropriate values of their utility functions when making choices. If this were the case, however, it would be hard to understand how people could be subject to preference reversals and other inconsistencies in revealed preferences.

The view adopted by several researchers is that, instead of consulting existing preference functions, people choose by constructing preferences in light of the tasks with which they are confronted (Slovic, Griffin, & Tversky [97]; Payne, Bettman, & Johnson [81]). The notion is not that all preferences are constructed from nothing but rather that people only have quite general notions of their preferences, e.g., although they prefer more to less money and less to more effort, they do not have precise representations of their hypothetical utility functions and so can not execute precise tradeoffs. Therefore, when confronting specific tasks that require precise answers (consider, e.g., willingness-to-pay judgments) the precision of their preferences will be affected by normatively irrelevant considerations such as the kinds of scales on which they are required to respond or the presence/absence of comparison points (Hsee [55]).

The position taken here makes much sense and it is perhaps surprising that researchers took so long to reach it. Simply stated, it gives meaning to the notion that people have preferences that, depending on their experience, can be described as varying in looseness. If one's preferences are loose and have to be expressed precisely in specific circumstances, irrelevant task effects should be expected to occur. If, on the other hand, preferences are not loose, they should be relatively impervious to how they are elicited. Thus, experience in specific domains should lead to less "looseness" and greater invariance to irrelevant factors. At the same time, one might expect financial incentives to have the same kinds of effects but this has not always been found to be the case (Camerer & Hogarth [9]).

Related to this topic is the important finding that people seek to avoid making tradeoffs (see, e.g., Luce, Payne, & Bettman [74]). There are two reasons: first, tradeoffs can be hard to execute from a cognitive viewpoint; and second, making tradeoffs is difficult emotionally. This raises the issues of how much people really lose by avoiding tradeoffs (in terms of making optimal decisions) and whether they can ever learn to make certain kinds of tradeoffs. To be able answer both of these questions, however, one would need an extensive ecological analysis.

10. The role of emotions, affect, and intuition. As noted above, the study of judgment and decision making owes much to formal models of decision making in economics and statistics (e.g., expected utility and subjectively expected utility) as well as the cognitive revolution in psychology. Thus the emphasis of most work across the last five decades has been based on cognitive explanations with little apparent interest in the topics of emotion, affect, and intuition. Thus, for example, several key papers demonstrated how cognitive explanations could account for phenomena that were previously thought to be motivational in origin (see e.g., Brehmer [3], for why conflicts are sometimes hard to resolve, and Einhorn & Hogarth [24], for biases in learning).

Recent years, however, have seen increasing recognition of the importance of emotions, affect, and intuition in decision making. It is thus illuminating to trace the origins of this trend.

In 1968, Robert Zajonc published a remarkable paper in which he documented a phenomenon known as the mere exposure effect, the finding that people acquire positive affect to stimuli through mere exposure. His observations started by noting that people seem to have greater positive affect for more frequently occurring objects. He illustrated this by examining words. For example, he presented people with lists of antonyms, that is, pairs of words that have opposite meanings such as forward/backward, high/low, or on/off. He then asked his subjects to state which word in each pair had "the more favorable meaning, represented the more desirable object, even, state of affairs, characteristic, etc." (Zajonc [106], p. 5) Zajonc also determined the frequency with which each word was used in English. His results showed a stunning relation between preferences and relative frequencies. For example, at least 97 percent of his subjects preferred forward to backward, high to low, and on to off. In English, the first of each of these words appears more frequently than the second. For example, forward is approximately 5.4 times more frequent than backward. The analogous figures for the other pairs are 1.4 and 8.3, respectively[5].

Zajonc went on to provide further demonstrations of these effects using fake words said to be Turkish, Chinese-like characters, and photographs of students. It was further demonstrated that even when people's ability to identify stimuli is degraded to chance levels, positive affect is also a function of mere exposure (Kunst Wilson & Zajonc [67]).

Why does this affect occur? Noting that none of the stimuli investigated involve negative consequences, Zajonc proposed that increasing familiarity with stimuli that are not harmful induces positive affect through a simple process of learning. In other words, it is more important to learn when something new is potentially harmful or dangerous than when it is not. Thus, if something new is not perceived as negative, its default value is positive. Moreover, the more it is seen, the more positive affect is reinforced.

Zajonc's work has profound implications for how people acquire preferences, a topic that is clearly outside the scope of economics and that has hardly been touched by researchers in judgment and decision making. However, as simple casual observation can inform us, many firms bet much money in their advertising campaigns on the veracity of the mere exposure effect. Consider, in particular, advertisements that simply emphasize the name of a product or a firm without providing any information about particular products, e.g., Benetton or, say, Toshiba. Here the advertising strategy simply

[5] The frequencies are taken from the so-called L count of Thorndike and Lorge [98].

consists of reinforcing the familiarity of the brand name through repeated exposure.

I have often wondered why decision researchers have not pursued the implications of the mere exposure effect. In particular, the phenomenon implies that tacit learning is an important source of preferences and that one might be able to predict a person's preferences by understanding the experiences to which he or she has been exposed. One reason could be the lack of a normative model of preference updating to provide a benchmark. For instance, the existence of Bayes' theorem provides a clear benchmark for studying how people update beliefs and the documentation of systematic departures from the standard (see, e.g., Hogarth & Einhorn [51]). Lacking a counterpart for preferences, it is less obvious how to proceed[6].

Finally, building on his work on the mere exposure effect, in 1980 Zajonc challenged the then overtly cognitively-oriented decision researchers by suggesting that "preferences need no inferences" or that affect (acquired subconsciously through mere exposure) was often the driver of judgment and choice as opposed to more cognitive forces. As discussed below, Zajonc's work has gained significant attention.

In judgment and decision research, the concept of intuition — although not neglected — has often had negative connotations. The "intuitive predictions" of clinicians, for example, are often disparagingly said to be inferior to those of statistical models. Similarly, intuition is associated with processes that cause people to make errors in judgments under uncertainty (cf., Tversky & Kahneman [100]) as well as unwarranted confidence in judgment (Kahneman, Slovic, & Tversky [60]).

One notable exception, however, has been the work of Hammond who — building on Brunswik's [5] perceptual model — conceived of modes of judgment that can vary on a continuum from highly intuitive at one end to highly analytical at the other. In addition to this cognitive continuum, Hammond also postulated that tasks confronted in the environment vary on the extent to which they induce intuitive or analytical thought. Moreover, when modes of cognition coincide with demands of tasks, judgments tend to be better (using a correspondence criterion). In an intriguing study of judgments made by highway engineers, Hammond and his colleagues found supporting evidence for this position (Hammond, Hamm, Grassia, & Pearson [42]).

The ideas pioneered by Zajonc (see above) had increasing influence in social psychology in the 1980s and culminated in what are generally known as dual-process theories of thought (Chaiken & Trope [12]). The key idea here

[6] The notion that one might think of modeling the learning of preferences (and thus that they could change) raised some interesting discussions at the conference. For economists, preferences are typically assumed to be exogenous and fixed. Thus, it seems odd even to consider the acquisition of preferences and how these might change.

is that thought has components of which people are aware (conscious) and of which they are not fully aware (subconscious) and that people can process information in two distinct ways.

This distinction is clearly made by Seymour Epstein [28] who has distinguished between what he labels experiential and rational modes of thought. As Epstein states, the experiential system is driven by emotions, it is intuitive, and automatic. It

> is assumed to have a very long evolutionary history and to operate in non-human as well as human animals ... it is a crude system that automatically, rapidly, effortlessly, and efficiently processes information .... Although it represents events primarily concretely and imagistically, it is capable of generalization and abstraction through the use of prototypes, metaphors, scripts, and narratives. (Epstein [28], p. 715).

The rational system, on the other hand,

> is a deliberative, effortful, abstract system that operates primarily in the medium of language and has a very brief evolutionary history. It is capable of very high levels of abstraction and long-term delay of gratification. (Epstein [28], p. 715).

As evidence for these two modes Epstein points to the fact that, when emotions are aroused, people tend to eschew logical arguments, emotions may themselves be triggered by preconscious thoughts, and that there is an important difference between learning intuitively through experience and intellectually through instruction or analysis. He also notes that although people are capable of rational analysis they are still heavily influenced by story telling, the appeal of pictures over words or statistics, and superstitious thinking. In the experimental work conducted to support this position, he leans heavily on vignettes created by Tversky and Kahneman to show that people can recognize when they are using the different models of thought (Epstein, Lipson, Holstein, & Huh [29]) as well as some intriguing choice tasks (Kirkpatrick & Epstein [64]; Denes-Raj & Epstein [19]).

It is important to emphasize that Epstein (like Hammond) does not say that people reason only with either the experiential or rational mode. Typically, reasoning will start with the experiential mode and be modified by the rational, depending on circumstances. As an example, reconsider the two versions of the Wason task described above. In the abstract version, it is difficult to think through the task because (unless we are logicians) we have no appropriate model. Moreover, our rational thought system may not be up to solving the problem appropriately. For the version set in the bar, however, we do have experiences that provide a "script" (see Epstein quote above) for thinking through the task. This is much easier for us and we can reach the correct solution almost effortlessly.

Here I have given much credit to Hammond and Epstein because they were pioneers in promoting a more comprehensive view of judgment and decision making as involving more than just cognitive components. Similarly, Slovic and his collaborators have provided much evidence emphasizing the role of affect in judgment and decision making (particularly concerning the assessment of risk - see above). Indeed, Slovic and his colleagues have recently made a forceful argument that many judgments are driven by an "affect heuristic" (Slovic, Finucane, Peters, & MacGregor [95]).

Kahneman [58] has also come to accept that judgmental processes are the product of two systems of thought. Moreover, the intuitive component is not only large but quite effective. (Consider again the two versions of the Wason task discussed above.) What is fascinating is the fact that much of the interest in the field was stimulated by Tversky and Kahneman's original demonstrations that "intuitive" processes could lead people to make erroneous judgments. Tversky and Kahneman saw the errors as a means to understand the underlying processes of judgment. Many others saw the errors as a general indictment of human judgmental abilities, a viewpoint that is now being rectified.

3 Some Major Challenges

So much for the past! Let me now turn my attention to what I consider the major challenges facing the field today. These concern four areas: (1) making links to and incorporating knowledge from neuroscience; (2) developing methodologies that can generalize laboratory findings; (3) being more effective at helping people make decisions; and (4) collaboration with sister disciplines in the social sciences. A major theme uniting all these concerns is the need to develop, first, more comprehensive theories that integrate findings from different areas (e.g., recognizing that judgment and decision making involves affective as well as cognitive components) and, second, more explicit theories that specify the conditions under which certain main effects (e.g., loss aversion or ambiguity avoidance) do or do not obtain. This, in turn requires that more explicit attention be paid to understanding how behavior interacts with the different kinds of environments in which it occurs (cf., the third key "idea or finding" discussed above).

Neuroscience. If one considers just the last decade or so, there can be no doubt that the scientific community has seen remarkable advances in understanding how the brain functions. Whereas we cannot yet measure precisely what happens when people process information, a number of methods have revealed glimpses of what our descendants will be able to see more clearly.

The evidence comes from several sources: invasive studies of nonhumans such as rats and primates whose brains have much in common with humans (see, e.g., LeDoux [68, 69]); and non-invasive studies of humans. Some of these studies involve the opportunistic collection of data from people who have particular forms of brain lesions or who suffer from illnesses associated with specific parts of the brain (such that one can infer what capacities these people are missing, e.g., Damasio [15]); the others are studies that use brain imaging technologies such as the electro-encephalogram method (EEG), positron emission topography (PET) and functional magnetic resonance imaging (fMRI) (for a good review, see Camerer, Loewenstein, & Prelec [10]).

From my — admittedly cursory — review of studies in this area, the general picture that emerges is that the brain has many different systems and that different parts are specialized in different functions. The amygdala, for example is an important repository of emotional reactions that the so-called higher cortical levels of the brain may or may not be able to control. Different areas of the brain seem to be involved when people are thinking about losses as opposed to gains; areas stimulated in thinking about ambiguous probabilities are different from those where probabilities are known; familiar faces are matched with different patterns of activities than unfamiliar faces; and so on (Camerer et al., [10]).

At the present time, this research is very much at an exploratory stage and many findings can be related to the notions of dual-process theories of thought (see above). Moreover, as measurement techniques improve (and they will) I suspect that we will have a rich store of information to sift and interpret. It is very exciting and I expect we will have many surprises.

Developing methodologies. Most research on judgment and decision making is conducted with carefully designed experiments. Showing that experimental claims rule out alternative explanations is an important feature of most work that demonstrates a high level of internal validity. However, one of the most important results in the field is the importance of contextual effects (consider again the example of the Wason task considered above) and this naturally raises the question of external validity: Do the results demonstrated in the laboratory apply in naturally occurring settings?

At a conceptual level, I believe that we can say that this problem does have a good solution. Moreover, it is not new. It is the notion of representative design that was originally elaborated by Egon Brunswik [5, 6]. Simply stated, think of an experiment - that involves the interaction between participants and the tasks with which they are confronted - as a sample. To what extent does this sample generalize to the population in which we are interested? Clearly, this depends on establishing that the characteristics of both the participants and the tasks they face are representative of the population (e.g., can be thought of as random samples). There are, of course, many studies in which people have looked at different types of experimental participants, e.g., students and professionals. However, the major gap lies in the sampling of tasks.

In a recent review, Dhami, Hertwig, and Hoffrage [20] made a thorough review of the concept of representative design and difficulties that have been encountered in its use. As they argue, there is some interest in seeking to go beyond the laboratory and to study decision making where it occurs (see, e.g., Lipshitz, Klein, Orasanu, & Salas [72]). However, perhaps the greatest obstacle to the adoption of the principles of representative design by researchers is its absence from textbooks - the concepts are not taught to students.

A further point made by Dhami et al. is that the development of modern technologies has facilitated the ability to sample situations in a representative manner. Let me mention a couple. First, the availability of computers allows us to construct simulation models and virtual worlds that far exceed our dreams of a few decades ago. Clearly, experiments in such worlds are not the same as "real" experience. However, clever models can go a long way to simulating different environments. As existing examples, consider flight simulators for teaching pilots or aircraft traffic controllers (for a wonderful example, see Shanteau, Friel, Thomas,& Raacke [86]).

Second, there exists today a plethora of recording instruments that allow one to capture behavior as it occurs. For example, I recently used mobile telephones to send messages to people at random moments of the day (Hogarth [50]). On receiving the messages, respondents answered a small questionnaire concerning the decisions they were taking at those moments. In this way, I obtained a random sample of their decision behavior. Whereas, there was much that was "wrong" with this study, I believe it is indicative of the kinds of samples we will be able to collect as technology develops. (For a clever use of head-mounted video cameras, see Omodei, McLennan, & Wearing [79]).

Finally, it is clear that experimenters can also use the principles of representative design to collect data in the laboratory. Indeed, specific attempts to do just this have cast doubt concerning the generality of such well-established phenomena as overconfidence (Gigerenzer, Hoffrage, & Kleinbölting [33]), hindsight bias (Hoffrage, Hertwig, & Gigerenzer [45]) and the overestimation of low-probability risks (Hertwig, Barron, Weber, & Erev [43]). In particular, the latter work emphasizes the important distinction between how people typically face risks in laboratory tasks - where all elements are neatly summarized and described - compared to experiencing these as they do and do not occur across time. In the former, small probability events are made quite salient. In the latter, by definition, low probability events are rarely encountered.

Helping decision making. It is generally accepted that the technology of decision making — or decision analysis (see, e.g., Keeney & Raiffa [63]) — is much ahead of people's ability to apply it. In my view, this suggests two really important challenges. One is to develop explicit theories that will indicate when simple decision models — such as relying on only one variable or taking an average — should be used. Clearly, some work has already been started on this topic. Indeed, Gigerenzer et al. [34] suggested the metaphor of people possessing an "adaptive toolbox" from which decision making tools can be taken and used as required. What we need, however, is a more complete list of the tools as well as specification of the tasks for which they are and are not suited.

Second, if we accept the fact that most decision are taken intuitively, it seems important that we develop means to educate people's intuitions. In a recent book (Hogarth [48]) I elaborated on this theme and presented a framework for suggesting how it might be done. In addition, I am currently engaged in trying to understand the conditions under which people should trust their intuitions or analysis when making decisions (Hogarth [49]).

It is not methods or principles of decision making that our field lacks. It is how to bring these tools within reach of those who need them.

Collaboration with sister disciplines. Research on judgment and decision making has always been an interdisciplinary field. As interest in the field continues to grow, there will be a natural tendency for groups to split off and create their own sub-fields. I see this as a real threat to progress because it is at the intersection of approaches and traditions that innovations tend to occur (Campbell [11]). There will therefore be a need for some scholars to play bridging roles between sub-fields so that ideas can circulate throughout the whole field.

4 The Next 50 Years ...?

I hope that my discussion of the last fifty years - as well as current challenges - has conveyed some sense of the interest and excitement that this field has generated. Judgment and decision making are ubiquitous and important activities. Moreover, contrary to an idea that one should only study decisions that are important (or have economic consequences), I want to stress the importance of studying "small" decisions. There are two reasons. First, our life consists mainly of small decisions. Thus, even if each small decision we take only has small consequences, the aggregate consequences of the small decisions we face in our lifetimes is huge. Second, given that our decision making habits have been formed and automated by making small decisions, it is unlikely that these can be suppressed when we face large decisions. Indeed, such habits can have large impacts on important issues.

What can we expect to see in the next 50 years? Answering this question is a daunting challenge, so let me simply suggest the following (almost by way of a summary):

1. Closer links in our understanding of judgment and decision behavior between biology (specifically neuroscience) and psychological models. This, in turn, will influence how economists think about these issues.
2. Greater methodological sophistication and more rigorous standards for judging experimental results. Parenthetically, I see this as one of the benefits the field is already experiencing from its contacts with experimental economics.
3. Ingenuous uses of technology for capturing behavior in naturally occurring situations and/or in experimental laboratories.
4. Many surprises - what these will be, however, I cannot say

References

1. Balzer, W. K., Doherty, M. E., & O'Connor, R. (1989). Effects of cognitive feedback on performance. *Psychological Bulletin*, 106, 410-433.
2. Blattberg, R. C., & Hoch, S. J (1990). Database models and managerial intuition: 50% model + 50% manager. *Management Science*, 36, 887-899.
3. Brehmer, B. (1976). Social judgment theory and the analysis of interpersonal conflict. *Psychological Bulletin*, 83, 985-1003.
4. Bruner, J., Goodnow, J. J., & Austin, G. A. (1956). *A study of thinking.* New York: Wiley.
5. Brunswik, E. (1952). *Conceptual framework of psychology.* Chicago: University of Chicago Press.
6. Brunswik, E. (1956). *Perception and the representative design of experiments* (2nd ed.). Berkeley: University of California Press.
7. Camerer, C. F. (1981). General conditions for the success of bootstrapping models. *Organizational Behavior and Human Performance*, 27, 411-422.
8. Camerer, C. F., & Johnson, E. J. (1991). The process-performance paradox in expert judgment: How can the experts know so much and predict so badly? In K. A. Ericsson & J. Smith (Eds.), *Toward a general theory of expertise: Prospects and limits.* Cambridge: Cambridge University Press.
9. Camerer, C. F., & Hogarth, R. M. (1999). The effects of financial incentives in experiments: A review and capital-labor-production framework. *Journal of Risk and Uncertainty*, 19, 7-42.
10. Camerer, C. F., Loewenstein, G., & Prelec, D. (2004). Neuroeconomics: Why economics needs brains. *Scandinavian Journal of Economics*, 106 (3), 555-579.
11. Campbell, D. T. (1969). Ethnocentrism of disciplines and the fish-scale model of omniscience. In M. Sherif & C. W. Sherif (Eds.), *Interdisciplinary relationships in the social sciences* (pp. 328-348). Xenia, OH: Aldine Publishing.
12. Chaiken. S., & Trope, Y. (Eds.). (1999). *Dual-process theories in social psychology.* New York. The Guilford Press.
13. Cohen, L. J. (1981). Can human irrationality be experimentally demonstrated? *Behavioral and Brain Sciences*, 4, 317-370.
14. Conlisk, J. (1996). Why bounded rationality? *Journal of Economic Literature*, XXXIV (June), 669-700.
15. Damasio, A. R. (1964). *Descartes' error: Emotion, reason, and the human brain.* New York: Avon.
16. Dawes, R. M. (1979). The robust beauty of improper linear models. *American Psychologist*, 34, 571-582.
17. Dawes, R. M., & Corrigan, B. (1974). Linear models in decision making. *Psychological Bulletin*, 81, 95-106.
18. Dawes, R. M., Faust, D., & Meehl, P. E. Clinical versus actuarial judgment. *Science*, 243, 1668-1674.
19. Denes-Raj, V., & Epstein, S. (1994). Conflict between experiential and rational processing: When people behave against their better judgment. *Journal of Personality and Social Psychology*, 66, 819-829.
20. Dhami, M. K., Hertwig, R., & Hoffrage, U. (2004). The role of representative design in an ecological approach to cognition. *Psychological Bulletin*, 130(6), 959-988.
21. Edwards, W. (1954). The theory of decision making. *Psychological Bulletin*, 51, 380-417.

22. Edwards, W. (1968). Conservatism in human information processing. In B. Kleinmuntz (Ed.), *Formal representation in human judgment.* New York: Wiley.
23. Einhorn, H, J., & Hogarth, R. M (1975). Unit weighting schemes for decision making. *Organizational Behavior and Human Performance*, 13, 171-192.
24. Einhorn, H, J., & Hogarth, R. M. (1978). Confidence in judgment: Persistence of the illusion of validity. *Psychological Review*, 85, 395-416.
25. Einhorn, H, J., & Hogarth, R. M. (1986). Judging probable cause. *Psychological Bulletin*, 99, 3-19.
26. Einhorn, H. J., Kleinmuntz, D. N., & Kleinmuntz, B. (1979). Linear regression and process tracing models of judgment. *Psychological Review*, 86, 465-485.
27. Epstein, S. (1994). Integration of the cognitive and psychodynamic unconscious. *American Psychologist*, 49, 709-724.
28. Epstein, S., Lipson, A., Holstein, C., & Huh, E. (1992). Irrational reactions to negative outcomes: Evidence for two conceptual systems. *Journal of Personality and Social Psychology*, 62, 328-339.
29. Gigerenzer, G. (1996). On narrow norms and vague heuristics: A reply to Kahneman and Tversky. *Psychological Review*, 103, 592-596.
30. Gigerenzer, G., & Goldstein, D. G. (1996). Reasoning the fast and frugal way: Models of bounded rationality. *Psychological Review*, 103, 650-669.
31. Gigerenzer, G., & Hoffrage, U. (1995). Hoe to improve Bayesian reasoning without instruction: Frequency formats. *Psychological Review*, 102, 684-704.
32. Gigerenzer, G., Hoffrage, U., & Kleinbölting, H. (1991). Probabilistic mental models: A Brunswikian theory of confidence. *Psychological Review*, 98, 506-528.
33. Gigerenzer, G., & Hug, K. (1992). Domain specific reasoning: Social contracts, cheating, and perspective change. *Cognition*, 43, 127-171.
34. Gigerenzer, G., Todd, P. M., & the ABC Research Group. (1999). *Simple heuristics that make us smart.* New York: Oxford University Press.
35. Goldberg, L. R. (1970). Man versus model of man: A rationale, plus some evidence, for a method of improving on clinical inferences, *Psychological Bulletin*, 73, 422-432.
36. Goldstein, D. G., & Gigerenzer, G. (2002). Models of ecological rationality: The recognition heuristic. *Psychological Review*, 109 (1), 75-90
37. Grether, D. M., & Plott, C. R. (1979). Economic theory of choice and the preference reversal phenomenon. *American Economic Review*, 69 (4), 623-638.
38. Grove, W. M., Zald, D. H., Lebow, B. S., Snitz, B. E., Nelson, C. (2000). Clinical versus mechanical prediction: A meta-analysis. *Psychological Assessment*, 12, 19-30.
39. Hammond, K. R. (1955). Probabilistic functioning and the clinical method. *Psychological Review*, 62, 255-262.
40. Hammond, K. R. (1990). Functionalism and illusionism: Can integration be usefully achieved? In R. M. Hogarth (Ed.), *Insights in decision making: A tribute to Hillel J. Einhorn.* Chicago: University of Chicago Press.
41. Hammond, K. R. (1996). *Human judgment and social policy: Irreducible uncertainty, inevitable error, unavoidable injustice.* New York: Oxford University Press.
42. Hammond, K. R., Hamm, R. M., Grassia, J., & Pearson, T. (1987). Direct comparison of the efficacy of intuitive and analytical cognition in expert judgment. *IEEE Transactions on Systems, Man, and Cybernetics*, 17, 753-770.

43. Hertwig, R., Barron, G., Weber, E. U., & Erev, I. (2004). Decisions from experience and the effect of rare events in risky choice. *Psychological Science*, 15 (8), 534-539.
44. Hoffman, P. J. (1960). The paramorphic representation of human judgment. *Psychological Bulletin*, 57, 116-131.
45. Hoffrage, U., Hertwig, R., & Gigerenzer, G. (2000. Hindsight bias: A by-product of knowledge updating? *Journal of Experimental Psychology: Learning, Memory, and Cognition*, 26, 566-581.
46. Hogarth, R. M. (1981). Beyond discrete biases: Functional and dysfunctional consequences of judgmental heuristics. *Psychological Bulletin*, 90, 197-217.
47. Hogarth, R. M. (1993). Accounting for decisions and decisions for accounting. *Accounting, Organizations and Society*, 18 (5), 407-424.
48. Hogarth, R. M. (2001). *Educating intuition.* Chicago: University of Chicago Press.
49. Hogarth, R. M. (2005). Deciding analytically or trusting your intuition? The advantages and disadvantages of analytic and intuitive thought. In T. Betsch & S. Haberstroh (Eds.), *The routines of decision making.* Mahwah, NJ: Erlbaum.
50. Hogarth, R. M. (2006). Is confidence in decisions related to feedback? Evidence from random samples of real-world behavior. In K. Fiedler & P. Juslin (Eds.), *In the beginning there is a sample: Information sampling as a key to understand adaptive cognition.* Cambridge, UK: Cambridge University Press.
51. Hogarth, R. M., & Einhorn, H. J. (1992). Order effects in belief updating: The belief-adjustment model. *Cognitive Psychology*, 24, 1-55.
52. Hogarth, R. M., & Karelaia, N. (2005). Ignoring information in binary choice with continuous variables: When is less "more"? *Journal of Mathematical Psychology*, 49, 115-124.
53. Hogarth, R. M., & Karelaia, N. (2006). "Take-the-Best" and other simple strategies: Why and when they work "well" with binary cues. *Theory and Decision*, 61, 205-249.
54. Hogarth, R. M., & Karelaia, N. (2006). Regions of rationality: Maps for bounded agents. *Decision Analysis*, 3(3), 124-144.
55. Hsee, C. K. (1996). Attribute evaluability: Its implications for joint-separate evaluation reversals and beyond. *Organizational Behavior and Human Decision Processes*, 67 (3), 247-257.
56. Hsee, C. K., Loewenstein, G. F., Blount, S., & Bazerman, M. H. (1999). Preference reversals between joint and separate evaluations of options: A review and theoretical analysis. *Psychological Bulletin*, 125 (5), 576-590.
57. Johnson, E. J., & Goldstein, D. (2003). Do defaults save lives? *Science*, 302, 1338-1339.
58. Kahneman, D. (2003). Maps of bounded rationality: Psychology for behavioral economics. *American Economic Review*, 93 (5), 1449-1475.
59. Kahneman, D., & Frederick, S. (2002). Representativeness revisited: Attribute substitution in intuitive judgment. In T. Gilovitch, D. Griffin, & D. Kahneman (Eds.), *Heuristics and biases: The psychology of intuitive judgment.* Cambridge, UK: Cambridge University Press.
60. Kahneman, D., Slovic, P., & Tversky, A. (Eds.). (1982). *Judgment under uncertainty: Heuristics and biases.* New York: Cambridge University Press.
61. Kahneman, D., & Tversky, A. (1979). Prospect theory: An analysis of decision under risk. *Econometrica*, 47, 263-291.

62. Kahneman, D., & Tversky, A. (Eds.). (2000). *Choices, values, and frames.* New York: Cambridge University Press.
63. Keeney, R. L., & Raiffa, H. (1976). *Decisions with multiple objectives: Preferences and value tradeoffs.* New York: Wiley.
64. Kirkpatrick, L. A., & Epstein, S. (1992). Cognitive-experiential self-theory and subjective probability: Further evidence for two conceptual systems. *Journal of Personality and Social Psychology*, 63, 534-544.
65. Klayman, J., & Ha, Y.-W. (1987). Confirmation, disconfirmation, and information in hypothesis testing. *Psychological Review*, 94, 211-228.
66. Kleinmuntz, B. (1990). Why we still use our heads instead of formulas: Toward an integrative approach. *Psychological Bulletin*, 107, 296-310.
67. Kunst-Wilson, W. R., & Zajonc, R. B. (1980). Affective discrimination of stimuli that cannot be recognized. *Science*, 207, 557-558.
68. LeDoux, J. (1996). *The emotional brain: The mysterious underpinnings of emotional life.* New York: Simon & Schuster.
69. LeDoux, J. (2002). *Synaptic self: How our brains become who we are.* New York: Viking Penguin.
70. Libby, E. (1976). Man versus model of man: Some conflicting evidence. *Organizational Behavior and Human Performance*, 15, 1-12.
71. Lichtenstein, S., & Slovic, P. (1971). Reversals of preference between bids and choices in gambling decisions. *Journal of Experimental Psychology*, 89, 46-55.
72. Lipshitz, R., Klein, G, Orasanu, J., & Salas, E. (2001). Focus article: Taking stock of naturalistic decision making. *Journal of Behavioral Decision Making*, 14, 331-352.
73. Loewenstein, G. F., Weber, E. U., Hsee, C. K., & Welch, N. (2001). Risk as feelings. *Psychological Bulletin*, 127, 267-286.
74. Luce, M. F., Payne, J. W., & Bettman, J. R. (1999). Emotional trade-off difficulty and choice. *Journal of Marketing Research*, 36, 143-159.
75. McGill, A. L. (1989). Context effects in judgments of causation. *Journal of Personality and Social Psychology*, 57, 189-200.
76. Meehl, P. E. (1954). *Clinical versus statistical prediction.* Minneapolis: University of Minnesota Press
77. Nisbett, R. E., Fong, G. T., Lehman, D. R., & Cheng, P. W. (1987). Teaching reasoning. *Science*, 238, 625-631.
78. Nisbett, R. E., & Ross, L. (1980). *Human inference: Strategies and shortcomings of social judgment.* Englewood Cliffs, NJ: Prentice-Hall.
79. Omodei, M., McLennan, J., & Wearing, A. J. (2005). How expertise is applied in real-world dynamic environments: Head-mounted video and cued recall as a methodology for studying routines of decision making. In T. Betsch & S. Haberstroh (Eds.), *The routines of decision making.* Mahwah, NJ: Erlbaum.
80. Payne, J. W. (1982). Contingent decision behavior. *Psychological Bulletin*, 92, 382-402.
81. Payne, J. W., Bettman, J. R., & Johnson, E. J. (1992). Behavioral decision research: A constructive processing approach. *Annual Review of Psychology*, 43, 87-131.
82. Payne, J. W., Bettman, J. R., & Johnson, E. J. (1993). *The adaptive decision maker.* New York: Cambridge University Press.
83. Peterson, C. R., & Beach, L. R. (1967). Man as an intuitive statistician. *Psychological Bulletin*, 68, 29-46.

84. Savage, L. J. (1954). *The foundations of statistics.* New York: Wiley.
85. Sawyer, J. (1966). Measurement and prediction, clinical and statistical. *Psychological Bulletin*, 66, 178-200.
86. Shanteau, J., Friel, B. M., Thomas, R. P., & Raacke, J. (2005). Development of expertise in a dynamic decision-making environment. In T. Betsch & S. Haberstroh (Eds.), *The routines of decision making.* Mahwah, NJ: Erlbaum.
87. Siegel, S. (1957). Level of aspiration and decision making. *Psychological Review*, 64, 253-262.
88. Simon, H. A. (1955). A behavioral model of rational choice. *Quarterly Journal of Economics*, 69, 99-118.
89. Simon, H. A. (1956). Rational choice and the structure of the environment. *Psychological Review*, 63, 129-138.
90. Simon, H. A. (1957). *Models of man.* New York: Wiley.
91. Simon, H. A. (1978). Rationality as process and product of thought. *American Economic Review*, 68 (2), 1-16.
92. Simon, H. A. (1979). Information processing models of cognition. *Annual Review of Psychology*, 30, 363-396.
93. Slovic, P. (1987). Perception of risk. *Science*, 236, 280-285.
94. Slovic, P. (2000). *The perception of risk.* London: Earthscan.
95. Slovic, P., Finucane, M., Peters, E, & MacGregor, D. (2002). The affect heuristic. In T. Gilovich, D. Griffin, & D. Kahneman (Eds.), *Intuitive judgment: Heuristics and biases.* Cambridge, UK: Cambridge University Press.
96. Slovic, P., Fischhoff, B., & Lichtenstein, S. (1976). Cognitive processes and societal risk taking. In J. S. Carroll & J. W. Payne (Eds.), *Cognition and social behavior.* Potomac, MD: Lawrence Erlbaum Associates.
97. Slovic, P., Griffin, D., & Tversky, A. (1990). Compatibility effects in judgment and choice. In R. M. Hogarth (Ed.), *Insights in decision making: A tribute to Hillel J. Einhorn.* Chicago: The University of Chicago Press.
98. Thorndike, E. L., & Lorge, I. (1944). *The teacher's word book of 30,000 words.* New York: Teachers College, Columbia University.
99. Tversky, A., & Kahneman, D. (1974). Judgment under uncertainty: Heuristics and biases. *Science*, 185, 1124-1131.
100. Tversky, A., & Kahneman, D. (1983). Extensional versus intuitive reasoning: The conjunction fallacy in probability judgment. *Psychological Review*, 90, 293-315.
101. Tversky, A., & Kahneman, D. (1986). Rational choice and the framing of decisions. *Journal of Business*, 59 (4, Pt. 2), S251-S278.
102. von Neumann, J., & Morgenstern, O. (1947). *Theory of games and economic behavior* (2nd ed.). Princeton, NJ: Princeton University Press.
103. Wason, P. C. (1960). On the failure to eliminate hypotheses in a conceptual task. *Quarterly Journal of Experimental Psychology*, 12, 129-140.
104. Wason, P. C. (1966). Reasoning. In B. M. Foss (Ed.), *New horizons in psychology.* Harmondsworth: Penguin.
105. Yaniv, I., & Hogarth, R. M. (1993). Judgmental versus statistical prediction: Information asymmetry and combination rules. *Psychological Science*, 4, 58-62.
106. Zajonc, R. B. (1968). Attitudinal effects of mere exposure. *Journal of Personality and Social Psychology Supplement*, 9(2,Pt.2), 1-27.
107. Zajonc, R. B. (1980). Feeling and thinking: Preferences need no inferences. *American Psychologist*, 35, 151-175.

On the Weighting of Rare Events and the Economics of Small Decisions *

Ido Erev

Technion - Israel Institute of Technology

Summary. Experimental research suggests that decision makers tend to overweight low probability (rare) events when they rely on a description of the possible outcomes (e.g., the situations addressed by Kahneman & Tversky [19]), but to underweight low probability events when they rely on personal experience (e.g., Barron & Erev [3]). The current chapter summarizes two lines of research designed to evaluate the implications of this pattern. The first line includes an experimental examination of the two contradicting effects. The results suggest that the two effects do not cancel each other. Rather, it is possible to predict which effect is likely to occur in a particular situation. The second line of research explores if the understanding of the experimental results can be used to derive practical implications. Four examples are presented that demonstrate that the experimental pattern can shed light on the economics of small decisions.

Many natural activities involve small decisions. For example, writing involves selection of words and sentences, and reading involves selection among alternative interpretations of sequences of letters. Small decisions are defined here by two properties. First, these decisions are not likely to be highly consequential; that is, the expected attractiveness of the outcomes from the different alternatives tends to be similar. Second, while making small decisions the decision makers do not have in front of them a description of the problem, and have no ability and/or incentive to carefully consider the outcome distributions associated with the feasible alternatives. Thus, in small decisions the decision makers have to make decisions based on their prior experience, and not on an explicit description of the possible consequences.

---

* Please address all correspondence to Ido Erev, the Max Werthiemer Minerva Center for Cognitive Studies, Faculty of Industrial Engineering and Management, Technion, Haifa, Israel, erev@tx.technion.ac.il. This research was supported by a grant from USA-Israel Binational Science Foundation, and benefited from comments from Alvin E. Roth, Ernan Haruvy and the participants of the International Conference "Experiments in Economic Sciences: New Approaches to Solving Real-world Problems"Okayama, December 2004.

Notice that the fact that small decisions are likely to be inconsequential does not mean that they are *always* inconsequential, and/or that they are inconsequential in the aggregate. Unwise selection of sentences, for example, can annoy referees and be costly to researchers. Similarly, incorrect interpretation of warning signs can lead readers to take unnecessary risks. Particularly clear examples of nontrivial small decisions are made during driving. In these examples the outcomes include the (low probability) outcome "painful death" and the aggregate outcomes are highly consequential. For instance, in the USA the estimated cost of traffic accidents, triggered in part by ex post suboptimal small decisions is higher than 100 Billion dollars a year (see e.g., Blincoe [6]).

The current chapter reviews recent studies of the economics of small decisions. It focuses on research conducted at the Technion. The chapter includes two main parts. The first part summarizes laboratory research that examines the basic properties of small decisions that are made based on personal experience. This section suggests that decisions from experience are very different from decisions that are made based on precise descriptions of the possible payoff distributions. Whereas people exhibit high sensitivity to low probability events in decisions from description, they behave as if they underweight low probability events in decisions from experience. The second section highlights some of the practical implications of the study of small decisions.

1 Decisions From Experience and Their Limited Correspondence to Decisions From Description

Mainstream decision research has focused on the way people make decisions when they receive a precise description of the decision problem. For example, in one of the problems studied in Kahneman and Tversky's ([19], Problem 1d in Table 1) influential research, the participants were asked to select one of the following alternatives:

- S 3,000 with certainty
- R 4,000 with probability 0.8, 0 otherwise.

This simple "decisions from description" experimental paradigm has two desirable qualities. First, because it makes probabilities explicit, it allows direct evaluation of expected utility theory (von Neumann & Morgenstern [24]). Second, it conforms to our intuition about the critical stage of large decisions. That is, when a CEO has to make a critical decision, she may ask experts to estimate the relevant payoff distributions. The experts' task in this case is to analyze all the relevant information and summarize it with a list of the possible outcomes and their probabilities for each of the feasible alternatives. Once the CEO receives this summary, her decision is similar to the decision problems studied in basic decision research.

The main results of Kahneman and Tversky's [19] classical study of decisions from descriptions are summarized in the central columns of Table 1.

	Kahneman & Tversky [19] (1979) (one shot)		Barron & Erev [3] (repeated play)	
		P(S)		P(S)
Gain, large P	Problem 1d S 3000 with certainty R 4000 with p =0.8; 0	80 %	Problem 1e S 9 with certainty R 10 with p =0.9; 0	44 %
Gain, low P	Problem 2d S 5 with certainty R 5000 with p = .001; 0	28 %	Problem 2e S 3 with certainty R 32 with p = 0.1; 0	72 %
Loss, large P	Problem 3d S -3000 with certainty R -4000 with p =0.8; 0	8 %	Problem 3e S -9 with certainty R -10 with p =0.9; 0	63 %
Loss, low P	Problem 4d S -5 with certainty R -5000 with p =.001; 0	83 %	Problem 4e S -3 with certainty R -32 with p =0.1; 0	40 %

Table 1. Percentage safe option choices (P(S)) as a function of payoff domain, and P - the probability of the gambles' extreme outcome. The left-hand column presents representative problems from Kahneman and Tversky's [19] study of decisions from experience. The right-hand column presents representative problems from Barron and Erev's [3] study of decisions for experience.

As noted by Kahneman and Tversky, this pattern implies a fourfold risk pattern. That is, the tendency to prefer the safer prospect appears to depend on an interaction between two factors, the payoff domain (gain or loss), and the magnitude of the probability of the most extreme (highest absolute value) outcome. They model decision makers as having a value function that is concave in gains and convex in losses, and a probability weighting function that overestimates small probabilities. Together, they report these produce risk aversion for gains and risk seeking for losses when the probabilities are moderate. But when the probabilities of gain or loss are small, the tendency to overweight small probabilities produces risk seeking for gains, and risk aversion for losses. For example:

Problem 1d, (3000 with certainty) or (4000 with probability .80; 0 otherwise), involves possible gains, and high probability for extreme outcome (80% to get 4000). In this setting most participants (80%) exhibit risk aversion: they prefer the safer prospect.

Problem 2d, (5 with certainty) or (5000 with probability 1/1000; 0 otherwise), involves possible gains and low probability for extreme outcome (1/1000 to get 5000). In this setting most participants (72%) exhibit risk seeking: they prefer the gamble.

Problem 3d, (-3000 with certainty) or (-4000 with probability .80; 0 otherwise), involves possible losses, and high probability for extreme outcome (80%

to lose 4000). In this setting most participants (92%) exhibit risk seeking: they prefer the riskier prospect.

Problem 4d, (-5 with certainty) or (-5000 with probability 1/1000; 0 otherwise), involves possible losses and low probability for extreme outcome (1/1000 to lose 5000). In this setting most choices (83%) exhibit risk aversion: they prefer the safer prospect.

Barron & Erev [3]; and see Hertwig et al. [17]) examine whether the behavioral regularities observed in the study of decisions from description are robust to situations in which decision makers have to rely on their personal experience of realized outcomes, rather than having explicit descriptions. To address these situations Barron and Erev used the minimal information experimental paradigm summarized in Figure 1. The participants were seated in front of a computer screen that presented two unmarked keys, and were told that their task is to operate a two-key money machine. The participants were told that the experiment includes many trials, and their task is to select one of the two keys in each trial. They were informed that each selection would result in a draw of a payoff from the selected key's payoff distribution. This payoff will be presented on the key and will be added to their total earnings. The participants were told that their goal is to maximize their earnings, but were told nothing about the payoff distributions. [1] Thus, they had to rely on the feedback they obtained after each choice.

Notice that the computation of the optimal or "rational" behavior in this simple paradigm is not trivial. That is, almost any behavior can be justified as rational based on certain assumptions concerning the decision maker's prior beliefs.

The right hand column in Table 1 highlights the pattern observed by Barron and Erev [3] in the study of decisions from experience. This pattern reflects a reversed fourfold risk pattern: Problem 1e, (9 with certainty) or (10 with probability .90; 0 otherwise), involves possible gains, and high probability for extreme outcome (90% to get 10). In this setting most choices (56% of the choices over the 200 experimental trials) exhibit risk seeking: they prefer the riskier prospect.

Problem 2e, (3 with certainty) or (32 with probability .1; 0 otherwise), involves possible gains and low probability for extreme outcome (0.1 to get 32). In this setting most choices (72%) exhibit risk seeking: the safer prospect was more popular.

---

[1] We also ran experiments in which the participants receive a precise description of the payoff distributions (see Perry, Haruvy & Erev [21]; Yechiam, Barron & Erev [28]). As noted below, this research leads to the surprising finding that the added information has little effect. In the context of decisions from experience people behave as if they do not read and/or do not pay attention to the instructions. As noted by Vernon Smith, this behavior is not unique to experimental subjects. Many of us exhibit similar behavior when we buy a new computer.

Instructions:

In this experiment you are operating a money machine. Upon pressing a button, you will win or lose a number of points. Your goal is to complete the experiment with as many points as possible. It is given that there is a differ ence between the buttons. The basic payment is 5 sheqels. Your final payment is comprised of the points you earn (200 points = 10 agorot) and the basic payment. For your information, the exact "machine" is likely to differ between participants.

Good luck.

The experimental screen:

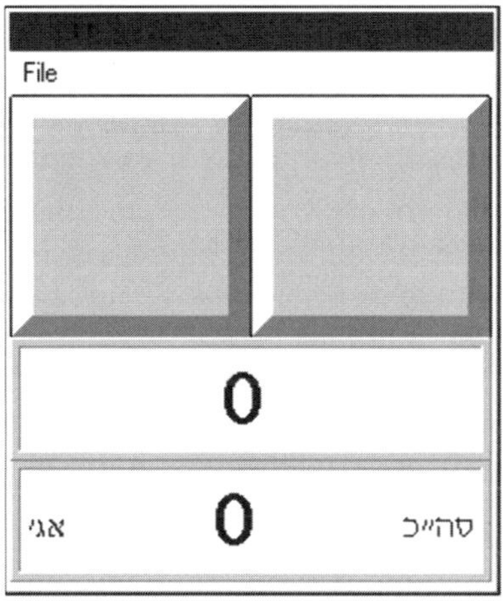

Note: The Hebrew words on the screen are "agorot" (4 agorot = 1 US cent) and "total."

Fig. 1. The instructions and the experimental screen in one of the conditions studied by Barron & Frey [3]

Problem 3e, (-9 with certainty) or (-10 with probability .90; 0 otherwise), involves possible losses, and high probability for extreme outcome (90% to lose 10). In this setting most choices (63%) exhibit risk aversion: The safer prospect was more popular.

Problem 4e, (-3 with certainty) or (-32 with probability 0.10; 0 otherwise), involves possible losses and low probability for extreme outcome (0.1 to lose 32). In this setting most choices (60%) exhibit risk seeking: The risky prospect was more popular.

Analysis of all eight problems summarized in Table 1 reflects an apparently complex three-way interaction. That is, the observed risk attitude is highly sensitive to the exact combination of three factors: the payoff domain, the magnitude of the probability, and the experimental paradigm. A focus on the effect of the probability dimension reveals, however, that the results can be simply summarized as a product of a main effect of the experimental paradigm. In particular, Kahnman and Tversky's results imply that in decisions from description people exhibit oversensitivity to low probability events, and Barron and Erev results imply that the opposite pattern emerges in decisions from experience. In decisions from experience decision makers tend to *underweight* rare events.

1.1 A Pessimistic and an Optimistic Interpretation of the Basic Experimental Results

Under a pessimistic interpretation of the experimental results, summarized above, the large difference between the two experimental paradigms implies that the observed regularities cannot be very robust. It can be argued that most natural decision problems are likely to fall between the two abstract experimental paradigms. Under this pessimistic assertion, the fact that the two paradigms lead to contradicting behaviors, implies that it is hard to predict which pattern will emerge in each situation. Indeed, it is possible that the two patterns will cancel each other.

Under an optimistic interpretation, the difference between the two paradigms is not a shortcoming. Rather, the coexistence of the different behavioral regularities highlights the importance of the interaction between the decision maker and the environment. It implies that a good understanding of the conditions that give rise to the different patterns can be used to derive useful predictions.

Some support for the optimistic interpretation is provided by previous research that illustrates the robustness of the different patterns. The robustness of the tendency to overweight low probability outcomes is clarified by the observation that many people buy both insurance and lotteries (see Friedman & Savage [13]). As noted by Kahneman and Tversky [19] the attractiveness of lotteries is consistent with the assertion that the low probability of winning is overweighted, and the popularity of insurance is consistent with the assertion that the low probability of major losses is overweighted. This explanation appears to outperform Friedman and Savage [13] assertion of an S-shape utility function, which is concave at low wealth levels (reflecting risk aversion) and convex at higher wealth levels (reflecting risk seeking).

Indications for the robustness of the tendency to underweight low probability events come from a wide set of experimental studies that used distinct experimental paradigms. One set of studies focused on the evaluation of signal detection theory [15]. This theory is a generalization of expected utility theory to the context of categorization and perceptual decisions. In a typical study participants are presented with a signal x (e.g., the height of a particular person) that was drawn from one of two distributions (states of nature). The participants' task is to detect (guess) the true state of nature (e.g., whether the person is male of female). Assume that the target state (e.g., Male) is a-priori less likely. The possible outcomes are typically denoted as: Hit if rare state is correctly detected; False alarm (FA) when the decision maker incorrectly states that the rare state occurred; Correct Rejection (CR) when the decision maker correctly notes that the rare state did not occur, and Miss otherwise. The guess implies a choice between two gambles:

"Guess Rare": {Hit with P(Rare|x); FA otherwise}
"Guess Freq": {CR with P(Freq|x); Miss otherwise}

where Hit, FA, CR, and Miss are the possible payoffs. The participants do not receive direct information concerning P(Rare|x), and P(Freq|x), but they are told or can learn the prior probability – P(Rare), and the likelihood ratios P(x|Rare)/P(x|Freq). Thus, in theory they can use Bayes rule to compute the gambles' probabilities. Optimal response requires equally sensitivity to the prior ratio: P(Rare)/[1-p(Rare)] and the payoff ratio: [Hit-FA]/[CR-Miss]. Experimental study of this prediction (see Barkan, Zohar & Erev [2]; and review in Erev [7]) shows higher sensitivity to the prior ratio than to the payoff ratio. This deviation from the optimal rule implies underweighting of rare events.

Weber et al. [26] and Hertwig et al. [17] show that the tendency to underweight outcome that occur with small probabilities emerge in a situations in which people have to rely on experience to make a single choice. In their studies the decision makers were asked to select once between two unknown payoff distributions. To form their preference the decision makers were allowed to sample the two distributions. The results show the pattern observed by Barron and Erev. The decision makers behave as if they rely on recent draws and underweight outcomes that occur with small probability.

Yechiam, Barron, and Erev [28] explored if the tendency to underweight low probability outcomes emerges when the payoff distributions are precisely presented before the decision maker. In their control condition (Condition Description 100) the participants were given the following instructions: "Your payoff in this experiment will be 2000 Agorot (showing up fee) [4.5 agorot were equal about 1 US cent] minus your loses during the experiment. Loses will be accumulated during 100 trials. In each trial you will play a gamble with negative outcomes. The outcomes of the gambles will be determined by the color (Red or Yellow) independently selected by the computer in each trial. Please indicate the number of trials that you want the computer to play each gamble (the total should be 100 choices)".

S	R
Lose 8 agorot if Red occurs (p = 0.005)	Lose 200 agorot if Red occurs (p = 0.005)
Lose 2 agorot if Yellow occurs (p = 0.995)	Lose 1 agora if Yellow occurs (p = 0.995)

In the Experience condition participants were given the following Instructions: "Your payoff in this experiment will be 2000 Agorot (showing up fee) minus your loses during the experiment. Loses will be accumulated during 100 trials. In each trial you will have to select a gamble (by clicking on it). The outcomes of the gambles will be determined by the color (Red or Yellow) selected by the computer after you make your choice". The results reveal more choices of the safer prospect in the control condition (60%) than in the experimental condition (30%). That is, the opportunity to rely of recent experiences led the participants to behave as if they provide less weight to the low probability Red event.

A direct indication of the robustness of the difference between decisions from descriptions and decisions from experience comes from a focus on the tendency to buy and use safety devices (see [3]). This research shows that in many cases people plan to use safety devices, and tend to use them initially. Yet with experience they behave as if they learn that these devices are not useful. For example, in 2001 most Israelis who bought car radio were willing to pay more in order to get a radio with a detachable panel. Most buyers detached the panel in the first few weeks, but stop detaching after few months.

1.2 Possible Explanations

Kahneman and Tversky [19] capture the tendency to overweight low probability outcomes with prospect theory's weighting function. This function implies that the subjective weighting of these outcomes is larger than the objective probabilities. Under one interpretation of this assertion, the subjective weighting is a perceptual phenomenon (see Birnbaum & Navarrete [5]) that occur in decisions from description. People tend to pay attention to all the described outcomes. The tendency to pay attention to a low probability outcome, resembles the white bear effect (Wegner et al. [25]): When people are told not to think about White bear, they cannot stop thinking about it. Under this analogy, people cannot stop thinking about the extreme outcome even when they are told that the probability is very low.

Erev and Barron [9] note that the tendency to underweight rare events in decisions from experience is naturally captured as a tendency to rely on the small set of most recent past experiences (see related ideas in Kareev [20]). Rare events tend to be underrepresented in small samples. Thus, the tendency to select a best reply to past experience implies underweighting of rare events. Erev and Barron [9] shows that this idea can be captured in generalizations of Roth and Erev's ([22]; and see Erev & Roth [12]) reinforcement learning model. These learning models provide a good summary of the main experimental results, and allow useful prediction of decisions from experience in other settings.

In summary, the experimental results can be summarized with the assertion that low probability extreme outcomes receive too much attention when they are presented, but are forgotten when the decision makers can rely on recent personal experience (in which low probability outcomes will seldom have been realized).

2 Practical Implications

The suggestion that the difference between decisions from description and decisions from experience is robust implies that the understanding of this difference can be used to drive interesting practical implications. Under this hypothesis slight changes in the decision environment are likely to have large behavioral effects. The current section reviews four studies that support this hypothesis.

2.1 The Enforcement of Safety Rules

The current analysis has several implications for the design of safe working environments (see Erev & Rodensky [11]). First, rule enforcement is necessary even when safe behavior is the rational course of action. The explanation of the relevant risks might not be enough. When workers make decisions from experience they are likely to underweight the possible risk and behave as if they believe that it "won't happen to me."

A second implication concerns the effectiveness of rule enforcement systems in which a small proportion of violations are heavily punished (see Becker [3]). The current analysis implies that systems of this type are likely to be effective in the context of decisions from description, but ineffective in the context of decisions from experience. When decisions are made from experience the low probability punishments are likely to be underweighted.

A third implication is optimistic. It implies that the fact that workers take unnecessary risks and behave as if they ignore safety rules, does not imply they will object to attempts to enforce these rules. Indeed, the observation that low probability events are overweighted in decisions from description implies that when workers are explicitly asked to consider the safety issue they will agree that they want to behave safely, and will be happy to see that the management designs a rule enforcement system in order to help them achieve this goal.

The forth implication is relevant to environments, like the typical Israeli factory, in which the official punishment for violations of safety rules is rather large (e.g., a fine equal to a week's salary). Under the assumption that large punishments are expensive (e.g. in worker morale) and cannot be applied except rarely, the arguments presented above lead to the conclusion that a change in the enforcement system that facilitates lighter (gentler) punishments with higher probability can be more effective.

Erev and Rodensky [11] applied this approach in six Israeli factories. Each of these "intervention studies" included four main steps. Initially, there was an informal conversation with the senior manager in each company. At this meeting the researcher explained the argument presented above, and obtained the manager's agreement to a change in the rule enforcement system that facilitates gentle high probability punishments. In addition, one member of the company senior management team was assigned to coordinate the intervention. The second step involved designing the details of the intervention jointly with the coordinator, followed by a presentation to senior and middle management. The basic idea behind the change was the design of a mechanism by which supervisors will be encouraged to approach each worker who violates the safety rule and remind him that this behavior might result in injury, and will be recorded (if repeated). The official role of these "violations records" was to allow the management to positively reinforce workers who observe the safety rule by giving these workers a higher probability of winning a lottery. In addition the records were used to detect workers and supervisors who ignore the safety rules. Baseline data were collected during the third step, conducted about two months prior to intervention. The data included objective measures of the workers' safety behaviors (c.f. Figure 2). The actual intervention was conducted during the fourth step. The intervention started with a formal presentation of the new policy to all the workers. Figure 2 presents measures of safety related behavior before and after the presentation in one of the departments in one of the six factories. The data were collected by the research team, and were independent of the supervisors' comments and records.

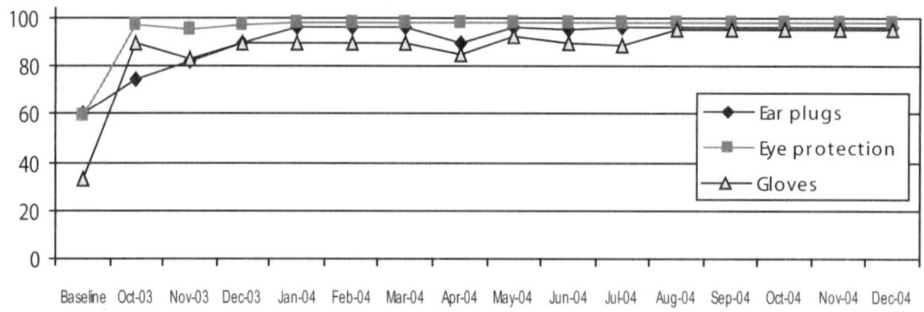

Fig. 2. Percentage of workers that obey the safety rule and use the required safety equipment as a function of time in one of the departments studied by Erev and Rodensky [11]. The baseline data were collected a month before the beginning of the intervention (in September 2003).

As demonstrated in Figure 2, the intervention had a large and immediate effect. A similar pattern was observed in all six factories. The rate of safe behavior increased to 90% immediately after the beginning of the intervention. More interesting is the observation that the effect of the intervention did not diminish with time. The rate of safe behavior increased or stayed high during the two years since the beginning of the intervention. Given the success of the intervention, and its relatively low cost, the factories have decided to maintain the experimental policy after the experiment.

2.2 Cheating in Exams

One of the possible explanations for the long term success of the gentle rule enforcement procedure, described above, is that rule enforcement problems tend to have two extreme equilibria. In one equilibrium, observing the rule is the norm, and the enforcers can easily detect and punish deviations (if they occur). Thus, no one is motivated to start violating the rule. In a second equilibrium violation is the norm, and the enforcers cannot address the frequent violations and, for that reason, violation is reinforcing. The possibility of two extreme equilibria and the hypothesis that small decisions are made based on recent experience implies that the effectiveness of different rule enforcement policies is likely to be particularly sensitive to the initial actions. Wise allocation of initial resources can lead to a convergence to the "good" equilibrium in which observing the rule is the norm.

Erev, Ingram, Raz and Shany [10] tried to apply this logic to reduce cheating during university exams. Their experiment was conducted during final semester exams of undergraduate courses at the Technion. Traditionally, instructions for exam proctors at the Technion included the following points:

(1) The student's ID should be collected at the beginning of the exam,
(2) A map of students' seating should be prepared.

Since the collection of the ID is the first step in the construction of the map, the common interpretation of these instructions was that the map should be prepared at the beginning of the exam. Early preparation of the map reflects an attempt to follow Becker's idea (preparing evidence to facilitate large punishments) but distracts the proctors, and reduces the probability of punishments (e.g., warning and/or writing the name of students who appear to cheat) at the beginning of the exam.

In the experimental conditions, to allow proctors to promptly warn students whose gaze was wandering, the second instruction to proctors was changed to state:

(2e) "A map of the students seating should be prepared 50 minutes after the beginning of the exam."

Seven undergraduate courses were selected to participate in the study. In all courses the final exam was conducted in two rooms. One room was randomly assigned to the experimental condition, and the second was assigned to the control condition. The only difference between the two conditions involved the timing of the preparation of the map in the instructions to the proctors. In the control group the instruction stated:

(2c) "A map of the students' seating should be prepared immediately after the beginning of the exam."

After finishing the exam, students were asked to complete a brief questionnaire in which they are asked to "rate the extent to which students cheated in this exam relative to other exams." A more direct measure of cheating will look at the correlation between seat proximity and responses (given an incorrect response).

The results reveal large and consistent difference between the two conditions. The perceived cheating level was lower in the experimental condition in all seven comparisons.

2.3 Two-stage Lotteries

The rule enforcement method discussed above is likely to be effective when the probability of punishment of each violation is high. This condition can be met when the rule enforcement agency can detect the violations with high probability, and can punish a large proportion of the detected violations. Perry, Erev and Haruvy [21] considered situations in which high probability detection is possible, but high probability punishment is not possible. Situations of this type are likely to emerge when detection can be made automatically (e.g., with a Red light camera), but the punishment requires a costly legal process.

Perry et al. note that the difference between decisions from experience and decisions from descriptions can be used to enhance the effectiveness of a rule enforcement mechanism in these "costly enforcement" environments. Specifically, they propose to present the punishment as a two-stage (bad) lottery. At the first stage, immediately after the violation of the rule, the violator is informed that the violation has been detected. At a second stage a small proportion of the violations is punished. Perry et al. show that when the second stage is delayed, this two-stage procedure can be rather effective.

2.4 The Effect of Rare Terrorist Attacks

Previous studies reveal that even rare terrorist attacks can have large negative effect on international tourism. For example, following the terrorist activity of the Sien Fein in Northern Ireland in the early 1970's, visitor arrivals fell from close to a million in 1967 to about 300,000 in 1976.

Yechiam, Barron & Erev [28] note that the current analysis implies that other effects of terrorism may not be as large. Specifically, it implies a large difference between international and local tourism. Traveling to a different country requires a big decision from description. Local tourism, on the other hand, can be a product of small decisions from experience (e.g., whether to take a sandwich to work or dine in a restaurant) and can be affected by experience. Thus, with experience the effect of rare terrorist attacks on local residents is likely to decrease.

Figure 3 presents the number of nights slept in Israeli hotels by local and international tourists before and after the beginning of the last wave of terrorist attacks in Israel (from September 2000). The results show a drop by both population with the beginning of the recent attacks, but a quick recovery by local tourists. This trend is consistent with the current predictions.

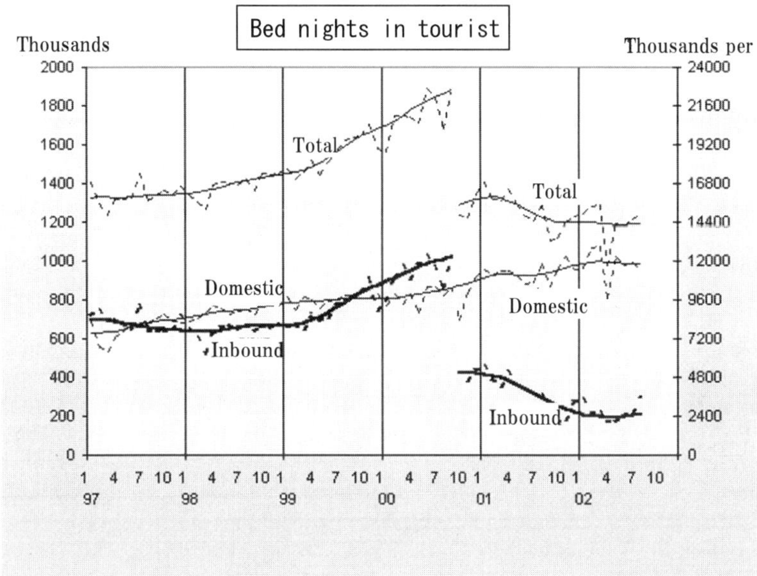

Fig. 3. Bed nights in tourist hotels in Israel from January 1997 to August 2002: seasonally adjusted average (dashed line) and trend by 1,000 bed nights (ICBS, 2002b. Used with permission).

3 Summary

The two lines of research, summarized above, suggest that the study of small decisions can be interesting.[2] The first line demonstrates that the characteristics of small decisions are not trivial. Repeated experience does not insure quick convergence of choice behavior toward maximization. Moreover, the deviations from maximization in small decisions from experience do not resemble the deviations from maximization in the well-studied paradigm of decisions

[2] Obviously, this suggestion does not imply the big decisions are less interesting. Paris Hilton highlights a particularly important big decision: The selection of the right parents (I first learned about this example from Amnon Rapoport). Interestingly, even this decision can be influenced by small decisions (the parents' behavior).

from description. A particularly important difference involves the weighting of rare events. Whereas people exhibit oversensitivity to low probability events in decisions from description, they behave as if they underweight low probability events in decisions from experience.

The second line of research demonstrates that the unique properties of small decisions are robust enough to facilitate the derivation of interesting practical implications.

References

1. Allais, M. (1953). Le comportement de l'homme rationel devant le risque, critique des postulates et axiomes de l'ecole americaine. *Econometrica*, 21, 503-546.
2. Barkan, R., Zohar, D., & Erev, I. (1998). Accidents and decision making under uncertainty: A comparison of four models. *Organizational Behavior and Human Decision Processes*, 74, 118-144.
3. Barron, G. & Erev, I. (2003). Small feedback based decisions and their limited correspondence to description based decisions. *Journal of Behavioral Decision Making*, 16, 215-233.
4. Becker, G.S. (1968). Crime and punishment: an economic approach. *Journal of Political Economy*, 76, 169-217.
5. Benartzi, S., & Thaler, R. (1995). Myopic loss aversion and the equity premium puzzle. *Quarterly Journal of Economics*, 110, 75-92.
6. Birnbaum, M. H., & Navarrete, J. B. (1998). Testing descriptive utility theories: Violations of stochastic dominance and cumulative independence. *Journal of Risk and Uncertainty*, 17, 49-78.
7. Blincoe, L.J. (1994). The Economic Cost of Motor Vehicle Crashes. *National Highway Traffic Safety Administration Technical Report.*
8. Erev, I. (1998). Signal detection by human observers: A cutoff reinforcement learning model of categorization decisions under uncertainty. *Psychological Review*, 105, 280-298.
9. Erev, I., Bereby-Meyer, Y., & Roth, A.E. (1999). The effect of adding a constant to all payoffs: experimental investigation, and implications for reinforcement learning models. *Journal of Economic Behavior and Organizations*, 39, 111-128.
10. Erev, I., & Barron, G. (2005). On adaptation, maximization, and reinforcement learning among cognitive strategies. *Psychological review.*
11. Erev, I., Ingram, P., Raz, O., & Shany, D., (2005). Organizational Design in the Face of Bounded Rationality: The potential for gentle rule enforcement. *Technion working paper.*
12. Erev, I. & Rodansky D. (2004). Gentle Enforcement of Safety Rules. A Final report of a research supported by the committee for accident prevention in the Israeli Ministry of Industry & Commerce (in Hebrew).
13. Erev, I., & Roth, A. (1998). Prediction how people play games: reinforcement learning in games with unique strategy equilibrium. *American Economic Review*, 88, 848-881.
14. Friedman, M., & Savage, L.J. (1948). The utility analysis of choice involving risk. *Journal of Political Economy*, 56, 279-304.

15. Green, D. M., & Swets, J. A. (1966/1988). Signal detection theory and psychophysics. New York: Wiley (reprinted in 1988; Los Altos, CA: Peninsula Publishers).
16. Grindle, A.C., Dickinson, A.M., & Boettcher, W. (2000). Behavioral safety research in manufacturing settings: A review of the literature. *Journal of Organizational Behavior Management*, 20, 29-68.
17. Haruvy, E., & Erev, I. (2002). On the application and interpretation of learning models. In R. Zwick, & A. Rapoport. (Eds.). *Advances in Experimental Business Research.* Boston MASS: Kluwer Academic Publishers.
18. Haruvy, E., Erev, I., & Sonsino, D. (2001). The medium prizes paradox: Evidence from a simulated casino. *Journal of Risk and Uncertainty*, 22, 251-261.
19. Hertwig, R., Barron, G., Weber, E. U., & Erev, I. (2004). Decisions from experience and the weighting of rare events. *Psychological Science.*
20. Herrnstein, R.J., Loewenstein, G.F., Prelec, D., & Vaughan, W., Jr. (1993). Utility Maximization and Melioration: Internalities in Individual Choice. *Journal of Behavioral Decision Making*, 6, 149-185.
21. Kahneman, D., & Tversky, A. (1979). Prospect theory: An analysis of decision under risk. *Econometrica*, 47, 263-291.
22. Kareev, Y. (2000). Seven (indeed, plus or minus two) and the detection of correlations. *Psychological Review*, 107, 397-402.
23. Perry, O., Haruvy, E., & Erev, I. (2002). Frequent Delayed Probabilistic Punishment in Law Enforcement. *Economics of Governance.* 3, 71-85.
24. Roth, A.E. & Erev, I. (1995), Learning in extensive form games: Experimental data and simple dynamic models in the intermediate term". *Games and Economics Behavior*, 8, 164-212.
25. Tversky, A., & Kahneman, D. (1992). Advances in prospect theory: cumulative representation of uncertainty. *Journal of Risk and Uncertainty*, 9, 195-230.
26. von Neumann, J., Morgenstern, O. (1947). *The Theory of Games and Economic Behavior*, Princeton, NJ: Princeton Univ. Press.
27. Wegner D. M., Schneider D. J. , Carter S. R. 3rd, White T. L. (1987) Paradoxical effects of thought suppression. *J. of Personality and Social Psychology.* 53(1):5 13.
28. Weber, E. U., Shafir, S., & Blais, A. (2004). Predicting risk sensitivity in humans and lower animals: Risk as variance or coefficient of variance. *Psychological Review.* 111, 430-445.
29. Yechiam, E., Barron, G., & Erev, I. (2005). The role of personal experience in contributing to different patterns of response to rare terrorist attacks. *J. of Conflict Resolution*, 49 (3), 430-439.

Trust, Fear, Reciprocity, and Altruism: Theory and Experiment

James C. Cox*

Georgia State University

Summary. This paper describes central topics in our research program on social preferences. The discussion covers experimental designs that discriminate among alternative components of preferences such as unconditional altruism, positive reciprocity, trust (in positive reciprocity), negative reciprocity, and fear (of negative reciprocity). The paper describes experimental data on effects of social distance and decision context on reciprocal behavior and male vs. female and group vs. individual differences in reciprocity. The exposition includes experimental designs that provide direct tests of alternative models of social preferences and summarizes implications of data for the models. The discussion reviews models of other-regarding preferences that are and are not conditional on others' revealed intentions and the implications of data for these models.

1 Introduction

The paper describes our research program on social preferences in which the central objective is to improve theory through a program of experimental testing and theoretical modeling motivated by data. Content will be drawn from several papers, and co-authors will be cited in the context of questions addressed by the research.

There are many other research programs in this area but I will only be discussing my own and my co-authors' research (with apologies to other researchers whose work is not discussed). The focus on our own research program facilitates a structured discussion of the relationship between specific features of experimental designs and theory development objectives.

This research program on social preferences is based on acceptance of the objective of parsimony in theoretical modeling (Samuelson [25]), of never

* Noah Langdale Jr Eminent Scholar Chair and Director of the Experimental Economics Center (ExCEN), Georgia State University. This paper was written while the author was a Visiting Scholar at the Workshop in Political Theory and Policy Analysis, Indiana University.

including within a model any complication that is not necessary to explain the phenomena being studied. Application of parsimony to theoretical modeling of social preferences and design of experiments to test the models is interpreted as leading to a focus first on the question of when the "economic man" model does not predict well, and models of "other-regarding" preferences are needed, and subsequently on when the other-regarding preferences need to include beliefs and/or intentions.

The logic of this application of parsimony is as follows. We begin by noting that the simplest model one can develop is the "economic man" model of self-regarding preferences in which the only thing an agent cares about in any context is his own material rewards. As is well documented by a large literature, the self-regarding preferences model does not predict well in many contexts in which distributional fairness is a salient concern.[1] This suggests that we first consider models of unconditional other-regarding preferences because such distributional preferences can be introduced into economic models by simply redefining the goods over which preferences are defined while preserving conventional regularity properties of the models such as completeness, transitivity, convexity and, perhaps, monotonicity. But if the other-regarding preferences that are modeled are conditional on — or "include" — reactions to others' past actions or beliefs about their future actions then that is a more fundamental departure from traditional economic theory. And so again, by the parsimony objective of theoretical modeling, one does not want to introduce such complications into models if, or when, they are not needed to maintain consistency between theory and data.

The approach of our research program is based on experimental designs that discriminate between the observable implications of unconditional other-regarding preferences and conditional preferences involving reactions to others' prior actions (such as positive or negative reciprocity) or beliefs about others' future actions (such as trust or fear). The reason to make these discriminations is that unconditional other-regarding preferences can be modeled without introducing intentions or beliefs by simply expanding the identity of goods to include other agents' incomes or consumption goods. In contrast, reciprocity makes preferences over goods dependent on perceptions of others' past actions (or attributions of their intentions) and trust or fear makes preferences over goods dependent on beliefs about others future reactions to one's own actions.

Another manifestation of application of the parsimony objective is that if experiments reveal that one needs to incorporate intentions and beliefs in some contexts, but not others, then it is desirable to develop a unified approach to modeling behavior in games both with and without reciprocal motivation. I shall discuss some models in the literature and our direct tests of those models, and then review some new models that my co-authors and I are developing for distinct patterns of behavior that are conditional, or are not conditional, on

---

[1] Fehr and Gächter [19] survey some of this literature.

others' revealed intentions. When discussing these models, I shall explain how they are variations on the same underlying model and hence an example of this last application of the parsimony objective, of developing a unified approach to modeling both less and more complicated instances of social preferences.

I begin with a game that provides an illustration of our approach to experimenting with fairness games. The game used in the discussion is the investment game.

2 An Example: The Investment Game

An experiment with the investment game has the following characteristics (see, for examples, Berg, et al. [2]; Cox [9]). Subjects are randomly paired. Each subject in each pair is given $10. Second movers are told to keep their $10. A first mover can either keep her $10 or give some or all of it to the second mover. Any amount given is multiplied by 3 by the experimenter. A second mover can either keep all of any amount received or return part or all of it to the paired first mover. The game is played only once. The experimental protocol uses double blind payoffs in which any individual subject's responses are anonymous to the experimenter and other subjects. All of the features of the experimental design and protocol are common information given to all subjects.

2.1 Predictions of the "Economic Man" Model

Predictions of the traditional economic man model for this game are transparent. Since second movers care only about their own material gain, they will keep any tripled amount sent by first movers. Since first movers care only about their own material gain, and know that second movers have the same kind of preferences, first movers will send nothing. Zero amounts returned and sent are the subgame perfect equilibrium of this game, given the economic man assumption about preferences. The predicted outcome is inefficient: each subject pair is predicted to get $20 in payoff — only the endowment — when it could have gotten as much as $40.

2.2 Behavior

Experiments with the investment game have been conducted by several researchers and the results look pretty much the same regardless of who runs the experiment. Figure 1 shows behavior in the investment game reported in Cox [9]. The amounts sent are represented by the striped bars and the amounts returned are portrayed by the cross-hatched bars. Of course, what the economic man model predicts is that there won't be any visible bars (of either type) in Figure 1. There are six subject pairs shown at the left side of the figure with no bars. The other 26 subject pairs do not behave like that. The overwhelming majority of first movers send money. Some second movers who receive money keep it all, as the economic man model predicts. So there are a few cases in which the first movers did not behave like economic man and the second movers did. But a large proportion of the second movers did not behave according to the economic man model either. There are even three subject pairs in which the first movers sent all \$10 and the second movers returned \$20, in other words the second movers chose the equal split fairness focal point in which each subject in the pair gets \$20, exactly double his/her endowment.

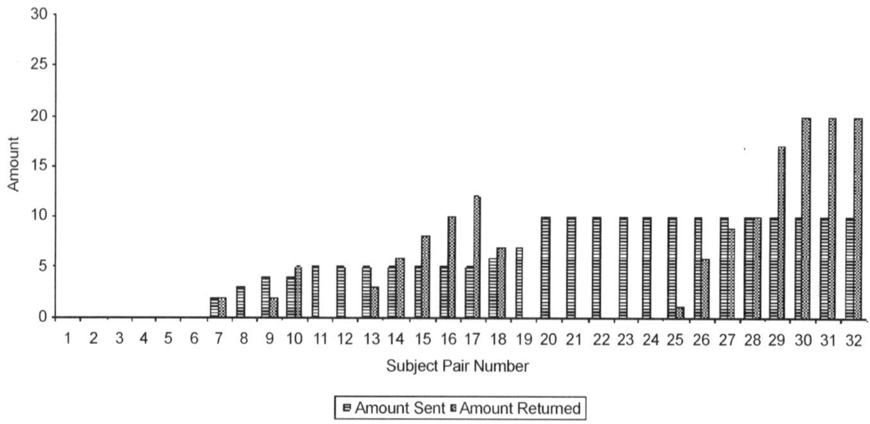

Fig. 1. Amounts Sent and Returned in Treatment A

Behavior in the investment game is representative of many games in the literature in which deviations from the economic man model's predictions are consistent with trust (by first movers) and reciprocity (by second movers). And many authors have concluded that trust and reciprocity have been observed in experimental games like this. But the experimental design actually does not support that conclusion. The reason is that first movers may send money to second movers because of unconditional altruistic preferences: it only costs a first mover 33 cents for each \$1 increase in the other person's money payoff. Furthermore, second movers may return money to first movers (who have less, after sending some of their endowment) because of either unconditional altruism or inequality aversion.[2]

[2] Models of inequality averse preferences are presented by Fehr and Schmidt [20] and Bolton and Ockenfels [3].

If this behavior could all be explained by unconditional altruism then that would be a relatively parsimonious extension of theory: just define the preferences over both my income and your income, assume positive monotonicity in both variables, and assume convexity of indifference curves. On the other hand, if subjects' behavior is characterized by trust and/or reciprocity then the implied changes in theory are less parsimonious. Modeling trust requires introduction of beliefs into theory. Modeling reciprocity requires introduction of perceived intentions into theory. These are more extensive and less tractable changes in theory than is modeling unconditional altruism and — according to the objective of parsimony — one does not want to introduce these complications into theory if they are not needed to explain behavior.

In order to proceed without ambiguity in discussing the relation between theory and alternative experimental designs, one needs some clearly-stated definitions of terms. Here are ones that will be used.

2.3 Definitions of Terms for Identifying Behavioral Motivations

Self-regarding (or "economic man") preferences are characterized by positively monotonic utility for one's own material payoffs and indifference about others' material payoffs. Other-regarding preferences are characterized by utility that is not constant with respect to variations in one's own or others' material payoffs. Altruistic preferences are characterized by utility that is monotonically increasing in others' material payoffs as well as one's own payoffs. Positive (direct) reciprocity is a motivation to adopt a generous action that benefits someone else, at one's own material cost, because that person's intentional behavior was perceived to be beneficial to oneself. Trust is a belief that one agent has about another. A trusting action is one that creates the possibility of mutual benefit and the risk of loss of one's own utility if the other person defects.

Why did I use the term "utility" in the definition of trust? Because if one is an altruist and would like to send some money to the other person, even if it was certain they wouldn't send anything back, then if the person doesn't, in fact, send anything back there may be no loss in utility. In that case, "trust" wouldn't be needed to explain the first mover's behavior; instead, the more parsimonious explanation — unconditional altruism — would suffice. Thus the question about identifying trusting behavior becomes: "Does a first mover in the investment game send more money to the second mover than he would in another game in which the first mover has the same set of feasible choices as in the investment game but knows for sure that the second mover cannot return anything?" This is clearly a different question than: "Does the first mover send any money to the second mover in the investment game?"

Negative (direct) reciprocity is a motivation to adopt an action that harms someone else, at one's own material cost, because that person' s intentional behavior was perceived to be harmful to oneself. Fear is a belief that one agent has about another. An action that is fearful of another is one that forgoes an otherwise preferred action because of a belief that the other agent will inflict costly punishment as a response to choice of the otherwise-preferred action. Negative reciprocity and fear will be discussed in section 3, in the context of experiments with the moonlighting game, but we first continue the discussion of experiments with the investment game.

2.4 Investment Game Triadic Design

Consider again the investment game, but instead of running it by itself, embed it in an experiment with three games, in what we call a "triadic design." Each game is an experimental treatment. The objective of the triadic experimental design is to construct treatments that reveal whether behavior in a central game of interest (in the present case, the investment game) can be represented with a model of unconditional other-regarding preferences or, instead, requires the less parsimonious approach of constructing a model that incorporates agents' attribution of the intentions revealed by others' past actions and/or beliefs about their future actions. In order to support observational discrimination between these distinct motives for behavior, dictator control treatments are designed so as to provide subjects with the same (own income, other's income) feasible choice sets as does the investment game but remove the decision opportunity of the paired subject, and thereby remove the possible effects of beliefs and intentions attribution on behavior.

The experimental design for the investment game includes the following three treatments. Treatment A is the investment game in which each first mover and each second mover is given a \$10 endowment and each \$1 increase in the second mover's money payoff costs the first mover 33 cents. Treatment B is a dictator game, with the same endowments as the investment game, which gives dictators the same feasible set of choices (over the ordered pairs of their own and the other's payoffs) that first movers have in the investment game. So what is the difference? First movers have exactly the same decisions to make, and the same feasible set, in treatment A and treatment B. The difference is that they know for sure that in treatment B the second movers cannot return anything. So if we observe that subjects send significantly less in treatment B than they do in treatment A, then we can conclude that amounts sent in treatment A cannot be fully explained by altruism, that we need something else, and the natural thing of course is trust. Why? Because in treatment A the first movers can trust that the second movers will share part of the increased total payoff from the tripling of amounts sent, and as a result send more in treatment A than in treatment B.

Treatment C is the dictator control treatment for reciprocity. Treatment C gives dictators the same choices and feasible sets that second movers have in the investment game. Treatment C is constructed as follows. The dictators correspond to the second movers in the investment game (treatment A). Of course, the non-dictators do not have a decision to make. Each dictator is given a \$10 endowment. Each non-dictator is given an endowment equal to the amount kept (i.e. not sent) by a specific first mover in treatment A. Furthermore, each dictator is given an additional dollar amount equal to the amount received by a specific second mover in treatment A from the tripled amount sent by a first mover in treatment A. The subjects are informed with a table of the exact inverse relation between the number of additional dollars received by a dictator and the endowment of the anonymously-paired other subject. The subjects are not informed that their endowments are determined by choices of subjects in treatment A to avoid suggesting indirect reciprocity towards other subjects.

2.5 Conclusions About Behavior

Figure 2 shows behavior in Treatments A and B in the experiment reported in Cox [9]. This experiment was run with a double-blind payoff protocol in which the responses by individual subjects are anonymous to both the other subjects and the experimenter. Comparing the amounts sent in treatments A and B, one observes that more subjects send zero in the first mover dictator control (Treatment B) than in the investment game. Furthermore, more subjects send half ($5) or all ($10) of their endowments in the investment game than in Treatment B. So there is indeed a quite noticeable difference. Several parametric and non-parametric tests of these data support the conclusion that behavior in the investment game is known to exhibit trust because first movers send significantly more in the investment game (Treatment A) than in the first mover dictator control treatment (Treatment B). Thus behavior in the investment game is known to exhibit trust because first movers send significantly more in the investment game than in the first-mover dictator control treatment.

Figure 3 shows data for Treatments A and C. If one looks at the difference between the bars representing data from Treatments A and C, one sees a lot more cross-hatched bars of greater height, which suggests that play in the investment game is characterized by positive reciprocity. Several parametric and non-parametric tests support the conclusion that behavior in the investment game does exhibit positive reciprocity because second movers return significantly more in the investment game (Treatment A) than in the second mover dictator control treatment (Treatment C). Thus, behavior in the investment game is known to exhibit positive reciprocity because second movers returned significantly more in the investment game than in the second mover dictator control treatment.

2.6 Implications for Theory

This experiment has several implications for theoretical modeling. The first is that consistency with behavior requires theory to incorporate altruistic other-regarding preferences. The reason is that the majority of subjects send positive amounts of money to another in the dictator control treatments. Furthermore,

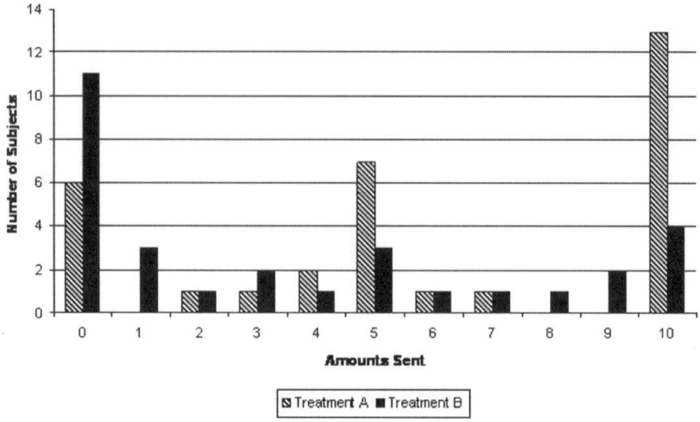

Fig. 2. Amounts Sent in Treatments A and B

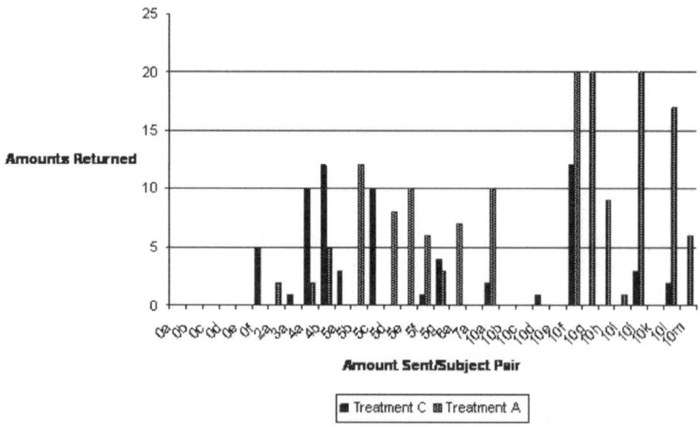

Fig. 3. Amount Returned in Treatments A and C

data-consistent theoretical models must incorporate beliefs about others' behavior because the triadic design reveals trusting behavior in the investment game. Finally, data-consistent models must incorporate other-regarding preferences that are conditional on the actions of others because the triadic design reveals positively reciprocal behavior in the investment game.

3 Conclusions From Other Experiments With Game Triads

The investment game is the first of several fairness games that my co-authors and I have experimented with using triadic designs. Another is the moonlighting game (Cox, Sadiraj, and Sadiraj [15]), which is an extension of the investment game in which both first and second movers can take money as well as give it.[3] Similarly to our finding for the investment game, we conclude that behavior in the moonlighting game exhibits both positive reciprocity and trust by comparing behavior in the central game of interest with behavior in appropriately-designed dictator control treatments. Unlike the investment game, the moonlighting game can elicit negative reciprocity and fear (of negative reciprocity) because subjects can take money from each other. Data from the moonlighting game and dictator controls provide weak support for negative reciprocity and fear because some test results are significant and others are not.

The trust game is a simplified version of the investment game in which the first mover can either "exit" (which corresponds to sending zero in the investment game) or "engage," that is to say move down the game tree, in which case the second mover has choice between keeping all of an increased total payoff or sharing it in one specific way.[4] Behavior in the trust game is invariant with a doubling of money payoffs (Cox and Deck [10]). Positive reciprocity is significant in the trust game with a single blind protocol but not with a double blind protocol (Cox and Deck [10]). This result has possible implications for understanding behavior in other games for which experimenters have only used single blind protocols.

In a single blind protocol, other subjects in an experiment cannot identify what a specific individual subject has done. In a double blind protocol, neither other subjects nor the experimenter can identify what any individual subject has done. Thus if the second mover, for example, wants to defect and keep all the money, that second mover does not have to worry about being frowned upon, or worse, perhaps not invited to be in future experiments or whatever else subjects might imagine, if the experimenter uses a double blind payoff protocol. In contrast, consider the implications of a single-blind protocol in a fairness game. For illustration consider the possible case in which a first mover has sent his entire \$10 endowment to the paired second mover. And suppose that the second mover considers keeping all of the \$40 and leaving the paired first mover with \$0. In a typical single blind protocol, the defecting second mover would be called by name to collect his \$40 in a face-to-face interaction with the experimenter. Furthermore, the experimenter is typically a professor, and a professor is arguably an authority figure for student subjects. The knowledge that subjects will have to face the experimenter to collect their

[3] The moonlighting game was introduced to the literature by Abbink, et al. [1].
[4] The trust game was introduced to the literature by McCabe and Smith [24].

payoffs does dissuade some potential defectors from defecting in the trust game. Since a large majority of experiments with fairness games have been run with single blind protocols, our finding may imply that some rethinking about conclusions is needed. One cannot know, a priori, all of the contexts in which a double blind protocol might yield different behavior than a single blind protocol. If one observes reciprocity in a double blind experiment then it is a really strong result which indicates that the norm for reciprocity is "internalized." In contrast, if reciprocity is observed in a single blind experiment, but not in an otherwise identical double blind experiment, then one needs to revisit the question of what the experimenter is attempting to measure because the experimenter-as-observer would have been shown to be a significant treatment. Furthermore, different experimenters may themselves have significant treatment effects: the prospect of collecting money payoffs resulting from defection from an old professor (arguably a father or mother figure) may be more constraining than the prospect of collecting such payoffs from a young graduate assistant.

Returning to results from experiments reported in our previous papers, one notes further conclusions as follows. We found that negative reciprocity and fear are not significant in the punishment mini-ultimatum game (Cox and Deck [10]), which is a simplified version of the traditional ultimatum game.[5] We found that play in the punishment mini-ultimatum game is invariant with framing the task as market exchange (Cox and Deck [10]). We also found that negative reciprocity is significant in the punishment mini-ultimatum game if it is embedded within a context of similar games but not when played in isolation (Cox and Deck [10]). This last finding is actually a little bit troubling for developing theoretical models in this area; it shows that it can indeed be quite a bit more complicated than we would like it to be. We also found that females are less positively reciprocal in investment games than are males, and that groups are less generous in the investment game than are individuals (Cox [7]). Cox and Deck [11] studies gender differences using a triadic experimental design including the trust game. The data indicate that women are more sensitive than men to the costs of generous actions. The factors that affect the level of observed generosity are reciprocal motivation, the level of money payoffs, and the level of social distance in the experimental protocol. The relatively greater sensitivity of women to the costs of generous behavior can explain much of the apparent inconsistencies among gender-difference experiments previously reported in the literature. Cox and Deck [12] reports a trust game with first mover trembling, which is a game in which "nature" randomly determines whether a first mover's decision is implemented or reversed. Data from this trust game with trembling indicate that second movers give first movers the benefit of the doubt in reacting to realization of the ungenerous branch of the game tree. However, first movers do not anticipate this forgiving

[5] Mini-ultimatum games were previously experimented with by Bolton and Zwick [4], Gale, et al. [21], and Falk, et al. [18].

response by second movers and are less likely to pursue the mutually beneficial outcome when there is trembling.

4 Models of Unconditional Other-regarding Preferences

Now I want to switch to my second theme and look at some specific models of social preferences. In models of inequality aversion, utility is increasing with one's own money payoff but decreasing with the difference between one's own and others' money payoffs.[6] In the quasi-maximin model, utility is increasing with an agent's own money payoff, with the lowest of all agents' payoffs (the maximin property), and with the total of all agents' payoffs (the efficiency property).[7] An alternative model motivated by data is the egocentric altruism model (Cox and Sadiraj [17]), which contains other-regarding preferences that are characterized by monotonicity, convexity, and egocentricity (defined below).

4.1 A Direct Test of Inequality Aversion

A direct test of inequality aversion is provided by the first-mover dictator control treatment for the investment game triad with the following design (Cox and Sadiraj [17]). The dictator is given \$10. The anonymously-paired subject is given \$10. The dictator can keep all of his \$10 or give any integral part of it to the paired person. Any amount given is tripled by the experimenter. Behavior in this experiment was as follows. First, 19 of 30 or 63% of the dictators gave positive amounts to the other person. The average amount given was \$3.63. The average payoff of dictators was \$6.37 and the average payoff of non-dictators was \$20.89, which implies a high degree of inequality favoring the other person. The inconsistency with the inequality aversion models does not just reflect an inconsistency with the parametric forms of these models; instead, it is a fundamental inconsistency with inequality aversion, per se. The behavior of the 37% of subjects that is consistent with inequality aversion is also consistent with self-regarding (or economic man) preferences, which is the preferred model because it is the simpler of the two. Therefore, inequality aversion is not needed to rationalize the behavior of any subjects in this experiment.

4.2 Direct Tests of the Quasi-Maximin Model

Cox and Sadiraj [17] report two direct tests of the quasi-maximin model with specially-designed dictator experiments. In each experiment, a dictator

6 See Fehr and Schmidt [20] and Bolton and Ockenfels [3].

7 The quasi-maximin model was introduced to the literature by Charness and Rabin [5].

is given a choice among three rows of a table containing payoffs for herself and three other people. In one experiment, the dictator's own payoff and the minimum payoff are the same in all three rows but the total payoff varies between rows. In the other experiment, the dictator's own payoff and the total payoff are the same in all three rows but the minimum payoff varies between rows. The experiment results are that the choices of 85% of the subjects in one experiment and of 94% of the subjects in the other experiment are inconsistent with quasi-maximin preferences.

4.3 More Information about Subjects' Preferences

The dictator experiment discussed in section 4.1, that provides a direct test of inequality aversion, reveals that a high majority of subjects behave like altruists when faced with the choice between choosing zero and giving a positive amount to the paired subject when the price of each $1 given is 33 cents. But this experiment leaves open the question of how subjects behave when they can either give or take money from another. Will they still appear to be altruists?

In experiment 1 of Cox and Sadiraj [17], a subject can choose zero or give money to the other subject or take money from him. The price of each $1 increase in the other subject's payoff is 33 cents. Each $1 taken from the other subject increases the dictator's payoff by $1. Thus the experiment reviewed here differs from the experiment reviewed in section 4.1 only by introduction of the opportunity to take money as well as give it or choose zero. This is the first-mover dictator control experiment for the moonlighting game (Cox, Sadiraj, and Sadiraj [15]). Data from this experiment are strikingly different from data for the experiment reviewed in section 4.1: the presence of the opportunity to take money causes a large majority of subjects to do just that; in fact, 69% of the subjects took money from the other person and 56% of them took the maximum possible amount of $5. Thus, in the absence of an opportunity to take money a high majority of subjects appear to be altruists but in the presence of opportunities to either give or take money a high majority of subjects appear to be selfish. Is this behavior contradictory?

4.4 The Egocentric Altruism Model

Behavior in these two dictator experiments can be rationalized by a model of other-regarding preferences with conventional properties known as the egocentric altruism model (Cox and Sadiraj [17]). A utility function $u(m, y)$ defined on the dictator's ("my") money payoff m and the paired subject's ("your") money payoff y that is monotonically increasing in both payoffs, has indifference curves that are strictly convex to the origin, and exhibits "egocentrism" can rationalize the data. Egocentrism is defined as $u(b, a) > u(a, b)$, for all a and b such that $b > a \geq 0$; in words, the individual is assumed to be an altruist but not a "Mother Teresa." An additional regularity property can be assumed for the model and maintain consistency with data described above: the utility function is assumed to be CES, hence homothetic, which implies that slopes of indifference curves are constant along rays from the origin, hence preferences over relative income m/y are defined in a straightforward way. The egocentric altruism model is consistent with almost all of the data from all four of the dictator experiments described in this section of the paper (Cox and Sadiraj [17]). Furthermore, this model is robust, it can also explain data from experiments with proposer competition and responder competition (Cox and Sadiraj [17]) and data from experiments with voluntary contributions to public goods (Cox and Sadiraj [16]).

5 Incorporating Intentions Into a Model of Social Preferences

As explained above, the egocentric altruism model can explain data from several types of dictator experiments while the inequality aversion and quasi-maximin models cannot explain such data. But neither the egocentric altruism model nor the other models incorporate intentions. Furthermore, this limitation is known to have empirical relevance because of experiments that identify the significance of reciprocity in various contexts, including experiments with the investment game (Cox [7], [9]), the trust game (Cox and Deck [10], [11]), and the moonlighting game (Cox, Sadiraj, and Sadiraj [15]).

An implication of the parsimony objective of theoretical modeling is that intentions should be incorporated into a model that can rationalize data from experiments without reciprocal motivation, such as dictator games, rather than proceeding in an orthogonal direction to develop unrelated models to explain intentions-conditional behavior such as reciprocity. This approach leads to development of a unified body of theory for modeling both less and more complicated instances of revealed social preferences.

5.1 A Parametric Model of Reciprocity and Fairness

The egocentric altruism model is extended to incorporate intentions in the "tractable model of reciprocity and fairness" (Cox, Friedman, and Gjerstad [13]) by assuming that the parameter weight that applies to another's money payoff is not an exogenous constant but, instead, is given by a function of a reciprocity variable r and a status variable s that are dependent on the other person's revealed intentions and social status characteristics that are relevant to the decision environment. The resulting utility function is a modified CES function of a decision-maker's own ("my") money payoff m and the other's ("your") money payoff y with a multiplicative weight for y given by the weighting function $\theta(r, s)$. The marginal willingness to pay to increase the other's payoff, when it is equal to one's own, is equal to $\theta(r, s)$. This θ function is assumed to be weakly increasing in both arguments, to have the neutral-state property given by $\theta(0,0) \geq 0$, and to be negative when r and s are sufficiently negative. Thus individuals are assumed to be non-malevolent in their baseline state of $(r,s) = (0,0)$, to be benevolent if the reciprocity and status variables are sufficiently positive, and to be malevolent if the reciprocity and status variables are sufficiently negative. In applying the model to data, the θ function is assumed to be identical across individuals except for a mean zero idiosyncratic term; in other words, individual agents are allowed to differ in their baseline altruism.

Data used in estimating the model come from several distinct types of experiments. Application of the model to data from the baseline dictator game, with random role assignment, reported by Cherry, Frykblom, and Shogren [6] yields estimates of individuals' residual or baseline altruism. Effects of experimenter-induced status on altruism are derived by applying the model to data from the dictator game with earned endowments reported by Cherry, Frykblom, and Shogren [6]. Estimates of subjects' reciprocity are derived by applying the model to data from Stackelberg duopoly (Huck, Muller, and Normann [23]) and moonlighting games (Cox, Sadiraj, and Sadiraj [15]). Reciprocity with context-dependent property rights is studied by applying the model to mini-ultimatum game data reported by Falk, Fehr, and Fischbacher [18]. Effects of reciprocity and status on subjects' other-regarding behavior are derived by applying the model to data from ultimatum games with random and contest assignment of the first-mover role, reported by Hoffman, McCabe, Shachat, and Smith [22]. Individuals' efficiency-increasing behavior is studied by applying the model to data from the first-mover dictator control treatment in the investment game triadic-design experiment reported in Cox [9].

5.2 A Non-parametric Model of Revealed Altruism

The egocentric altruism model and tractable model of reciprocity and fairness are further generalized in a non-parametric model based on partial orderings of preferences and opportunity sets (Cox, Freidman, and Sadiraj [14]). In this model, one agent's other-regarding preferences can depend on the actions of the other agent. The model is based on two partial orderings and two axioms that link them. The partial ordering of preferences is a formal representation of what it means for one preference ordering to be "more altruistic than" another. The partial ordering of opportunity sets is a formal representation of what it means for one opportunity set to be "more generous than" another. These two partial orderings are linked by two axioms. Axiom R states that more generous choices by the first mover in an extensive form game induce more altruistic preferences in the second mover. Axiom S states that the reciprocity effect on preferences is stronger following an act of commission by the first mover than following an act of omission. This non-parametric model is applied to data from investment and dictator games, carrot and stick games, Stackelberg duopoly game, and Stackelberg mini-game.

6 Summary

This research program involves experiments with "fairness games" designed to reveal the characteristics of individuals' social preferences and an approach to modeling these social preferences based on application of the objective of parsimony. The experiments reveal that behavior in fairness games exhibits unconditional altruism ("others' payoffs matter"), trust ("beliefs matter"), and reciprocity ("intentions matter"). Whether or not reciprocity is exhibited in some games depends upon whether the experimenter uses a single-blind or double-blind protocol ("who is observing matters") and the context in which a specific game is embedded ("fairness is a relative concept"). The experiments reveal that other-regarding behavior differs across small groups and individuals and across males and females ("type of decision-maker matters").

Modeling behavior in fairness games involves complications that vary with characteristics of the games. In simple dictator games that do not elicit reciprocal motivation, behavior is inconsistent with inequality aversion and quasi-maximin preferences. Behavior in these dictator games and in games of proposer competition, responder competition, and voluntary contributions to public goods can be rationalized by a model of egocentric altruism. Behavior in games such as the investment, trust, moonlighting, ultimatum, mini-ultimatum, Stackelberg duopoly, and Stackelberg mini-games that do elicit reciprocal motivation can be modeled with a tractable parametric extension of the egocentric altruism model and with a non-parametric revealed altruism model based on a partial ordering of preferences ("more altruistic than"), a partial ordering of opportunity sets ("more generous than"), and "reciprocity" and "status quo" axioms that link the two partial orderings.

References

1. Abbink, Klaus, Irlenbusch, Bernd, and Renner, Elke (2000), "The Moonlighting Game: An Empirical Study on Reciprocity and Retribution." *Journal of Economic Behavior and Organization*, 42, pp. 265-77.
2. Berg, Joyce, Dickhaut, John, and McCabe, Kevin (1995), "Trust, Reciprocity, and Social History," *Games and Economic Behavior*, 10, pp. 122-42.
3. Bolton, Gary E. and Ockenfels, Axel (2000), "ERC: A Theory of Equity, Reciprocity and Competition." *American Economic Review*, 90, pp. 166-93.
4. Bolton, Gary E. and Zwick, Rami (1995), "Anonymity versus Punishment in Ultimatum Bargaining," *Games and Economic Behavior*, 10, pp. 95-121.
5. Charness, Gary and Rabin, Matthew (2002), "Social Preferences: Some Simple Tests and a New Model," *Quarterly Journal of Economics*, 117, pp. 817-69.
6. Cherry, Todd L., Frykblom, Peter, and Shogren, Jason (2002), "Hardnose the Dictator," *American Economic Review*, 92, pp. 1218-21.
7. Cox, James C. (2002), "Trust, Reciprocity, and Other-Regarding Preferences: Groups vs. Individuals and Males vs. Females," in R. Zwick and A. Rapoport, (eds.), *Experimental Business Research*, vol. I, Kluwer Academic Publishers.

8. Cox, James C. (2003), "Trust and Reciprocity: Implications of Game Triads and Social Contexts," University of Arizona Discussion Paper Number 00-11.
9. Cox, James C. (2004), "How to Identify Trust and Reciprocity," *Games and Economic Behavior*, 46, no. 2, pp. 260-281
10. Cox, James C. and Deck, Cary A. (2005), "On the Nature of Reciprocal Motives," *Economic Inquiry*, 43, no. 3, pp. 623-635.
11. Cox, James C. and Deck, Cary A. (2006), "When are Women More Generous than Men?", *Economic Inquiry*, 44, no. 6, pp. 587-598.
12. Cox, James C. and Deck, Cary A. (2006), " Assigning Intentions when Actions are Unobservable: the Impact of Trembling in the Trust Game," *Southern Economic Journal*, 73, no. 2, pp. 307-314.
13. Cox, James C., Friedman, Daniel and Gjerstad, Steven (forthcoming), "A Tractable Model of Reciprocity and Fairness," *Games and Economic Behavior.*
14. Cox, James C., Friedman, Daniel, and Sadiraj, Vjollca (2006), "Revealed Altruism," Georgia State University, ExCEN working paper number 2006-09.
15. Cox, James C., Sadiraj, Klarita, and Sadiraj, Vjollca (forthcoming), "Implications of Trust, Fear, and Reciprocity for Modeling Economic Behavior," *Experimental Economics.*
16. Cox, James C. and Sadiraj, Vjollca (2007), "On Modeling Voluntary Contributions to Public Goods," *Public Finance Review*, 35, no. 2, pp. 311-332.
17. Cox, James C. and Sadiraj, Vjollca (2006), "Direct Tests of Models of Social Preferences and a New Model," Georgia State University, ExCEN working paper number 2006-13.
18. Falk, Armin., Fehr, Ernst, and Fischbacher, Urs (2003), "On the Nature of Fair Behavior." *Economic Inquiry*, 41, pp. 20-6.
19. Fehr, Ernst and Gächter, Simon (2000), "Fairness and Retaliation: The Economics of Reciprocity." *Journal of Economic Perspectives*, Summer, 14, pp. 159-81.
20. Fehr, Ernst and Schmidt, Klaus M. (1999), "A Theory of Fairness, Competition, and Cooperation," *Quarterly Journal of Economics*, 114, pp. 817-68.
21. Gale, John, Binmore, Kenneth G. and Samuelson, Larry (1995), "Learning to Be Imperfect: The Ultimatum Game," *Games and Economic Behavior*, 8, pp. 56-90.
22. Hoffman, Elizabeth, McCabe, Kevin, Shachat, Keith, and Smith, Vernon L. (1994), "Preferences, Property Rights, and Anonymity in Bargaining Games," *Games and Economic Behavior*, 7, pp. 346-80.
23. Huck, Steffen, Müller, Wieland, and Normann, Hans-Theo (2001), "Stackelberg Beats Cournot: On Collusion and Efficiency in Experimental Markets," *Economic Journal*, 111, pp. 749-66.
24. McCabe, Kevin A., and Smith, Vernon L. (2000), "A Comparison of Naive and Sophisticated Subject Behavior with Game Theoretic Predictions." *Proceedings of the National Academy of Sciences*, 97, pp. 3777-81.
25. Samuelson, Paul A. (1947), *Foundations of Economic Analysis* (Harvard University Press, Cambridge, MA).

What Have We Learned From Experimental Finance?

Shyam Sunder

Yale University

Summary. This paper addresses five questions about how stock market works and what we have learned from experiments in this field. 1, Why do we need even more data on financial market? Don't we have enough already? 2. How could the data from such small scale simple markets help us gain insights into far more complex investment environments? 3. Is experimental finance a branch or variation of behavioral economics/behavioral finance? 4. What have we learned so far from assets market experiments? 5. What is next?

1 Why More Data?

This question is frequently asked. Of all branches of economics, financial economics probably has available the most detailed and up-to-the-minute observational data from stock exchanges around the world. This branch of economics is characterized by a strong empirical tradition. Why, then, do we need to spend time and money to conduct experiments with financial markets and gather even more data?

Data from the stock exchanges include bids, asks, transaction prices, volume, etc. In addition, data from information services includes information on actions and events that may influence markets. Theories of financial markets (and economics of uncertainty more generally) are built on investor expectations. We need data on investor beliefs and expectations to empirically distinguish among competing theories. Yet, neither of these two sources of data does, nor can, report on investor expectations.

In experimental markets, the researcher knows the underlying parameters, and either knows or can make reasonable conjectures about the investor expectations. Armed with this knowledge, the researcher knows the price and other predictions of alternative theories. Indeed, the experiments are designed so the alternative theories yield mutually distinct predictions for the market. This approach allows us to conduct powerful tests of theories which are not possible from the field data alone; we know little about the parameters and expectations that generate the field data from stock exchanges. We shall return to illustrative examples in a later section after addressing the five questions.

2 What Can We Learn From Such Simple Markets?

Experimental markets are typically conducted in simple laboratory or classroom settings with a small number of student subjects who may have little prior experience with trading and investment. On the other hand, security markets are complex, populated by experienced sophisticated professionals. Naturally, the second question frequently asked is: What could we possibly learn from such Mickey Mouse experiments about the far more complex "real" markets? Experimenter may pay, say, \$50 to each participant after a two or three hour session while traders in the security markets we are interested in often have millions if not billions of dollars at stake.

All science is aimed at finding simple principles that explain or predict a large part (rarely all) of the phenomenon of interest. Simple models, whether in mathematics or in laboratory, make many assumptions. These can be divided into core assumptions and assumptions of convenience. Core assumptions are essential features of the environment while the convenience assumptions are made for mathematical tractability (e.g., probability distributions and preference functions in most cases). The power of a theory depends on the robustness of its explanations and predictions as the environments from which we gather the data deviate from the assumptions of convenience (Sunder [12]). The experimenter can deliberately and progressively make the experimental environment deviate from the assumptions of convenience in the theory to measure this robustness. This robustness check is not possible in the field data generated by the environment prevailing in the market.

In economics and finance, as in other sciences, simple experiments are used to discover and verify general basic principles. We learn to count through analogies of images or physical objects. We learn to swim in knee-deep waters. We learn and verify the laws of electricity, not by using a computer or radio, but by simple instruments such as potentiometer or ammeter. Manipulation of simple controls and monitoring the results builds the fundamental knowledge of science. The noise generated by countless factors in complex environments makes it difficult to detect the fundamental principles that might underline the economics of environment in which we are interested. Simple math and lab models help us learn, before we immerse ourselves in the complexity of the real world phenomena. If the principal is general, it should be applicable not only to complex but also to simple environments of a laboratory. If it does not pass the test of simple environments, its claim to be applicable to more complex environments is weakened.

3 Experimental Vs. Behavioral Finance

The third question often raised is: Is experimental finance the same as behavioral economics or behavior finance? My answer is no. In experimental economics the emphasis is on the design of markets and other economic institutions to gather empirical data to examine the consequences of institutions and rules. We assume that people do what they think is best for them, given what they think they know. They may be inexperienced, but they are not dumb; they learn. In experimental finance, we design experiments to sort out the claims of competing theories. On occasion we might conjecture new theory from the data but then we don't use the data used to generate the conjecture to test it. Like engineers, experimentalists design test beds to examine the properties and performance of alternative market institutions. The focus in this literature is on equilibrium, efficiency, prices and allocations. This work complements mathematical modeling and empirical research with field data.

4 What Have We Learned?

The fourth question is: What have we learned from experiments? Within the last couple of decades, asset market experiments have revealed some important findings by exploiting the advantages of laboratory controls. These findings were not and could not have been reached from field data or mathematical modeling alone. However, in combination with field data and modeling, laboratory experiments have helped us make substantial advances in our understanding of security markets. Let us review some key findings.

Security markets can aggregate and disseminate information. In other words, markets can be informationally efficient. However, just because they can doesn't mean they always are. Information dissemination, when it occurs, is rarely instantaneous or perfect; learning takes time. Efficiency is a matter of degree, not a 0-1 issue.

Plott and Sunder [7] asked if markets can disseminate information from those who know to those who don't. A satisfactory answer to this question could not be established from analysis of field data because we don't know which investor has what information. Plott and Sunder [7] used a simple experiment to address the question. As Table 1 shows, they designed a simple, single-period, two-state (X or Y) security, with the probability of each state given. The market was populated with four traders each of three types for a total of 12 traders; Type I received dividend of 400 in State X and 100 in State Y while the other two types received dividends of 300-150 and 125-175 respectively. Each trader was endowed with two securities and 10,000 in "cash" at the beginning of each period. The last column of Table 1 shows the expected dividends from the security for each of the three types of traders, when they do not know whether State X or Y is realized. Under this no information condition, the equilibrium price of the security would be 220, the maximum of the three expected values (220, 210 and 155) and Type I traders should buy all the securities at this price from the other traders.

Table 1. Information Dissemination Equilibria in a Simple Asset Market (Source: Plott and Sunder [7]

	State X Prob. = 0.4	State Y Prob. = 0.6	
Trader Type			Expected Dividend
I	400	100	220
II	300	150	210
III	125	175	155
PI Eq. Price Asset Holder	400 Trader Type I Informed	220 Trader Type I Uninformed	
RE Eq. Price Asset Holder	400 Trader Type I	175 Trader Type III	

Suppose the realized state is X and two traders of each type are informed at the beginning of the period that the state is X, and the other two are not. The informed traders know that the value of the dividend from the security (if they decide to hold it) is given in Column X, while the uninformed traders (assuming they are risk-neutral) would value the securities at the expected values given in the last column of the table. The equilibrium price would be the maximum of these six numbers 400, and Type I informed traders would buy the security at that price. If the realized state were Y instead, by a similar argument, the equilibrium price would be 220, the maximum of the six numbers in the Y and the expected value columns, and the Type I uninformed traders should buy the security at that price. This equilibrium is labeled Prior Information (PI) equilibrium because it assumes that the traders rely entirely on the information they receive at the beginning of the period, and do not learn any additional information about the realized state from their participation in the market.

PI equilibrium is problematic because it assumes that traders would not learn from their failures. Whenever Type I uninformed traders pay a price of 220 to buy a security, they will discover that the state turns out to be Y with a dividend of only 100, leaving them with a loss. If we assume that one cannot fool some of the people all the time, these traders should learn not to buy the securities at that price, making the PI equilibrium unsupportable.

Under the rational expectations (RE) or efficient market hypothesis, information about the state would be disseminated from the informed to the uninformed traders through the market process. Under this assumption, in

State X, all traders would know the state is X, will yield an equilibrium price of 400 which is the maximum of the three dividends in the column for State X, and all traders of Type I would buy the securities from the others. Similarly, in State Y, the equilibrium price would be 175 which is the maximum of the three dividends in the column for State Y, and all traders of Type III would buy the securities. This market was designed so the PI and the RE hypotheses yielded mutually distinct predictions of the market outcomes in prices and allocations.

Figure 1 shows the results for one of these markets. In Periods 1 and 2, traders were not given any information and the prices were located in the vicinity of the no information prediction of 220. In Period 3, State Y was realized, and the prices were much closer to the RE prediction of 175 than to the PI prediction of 220. Similar results were repeated in the other periods (5, 6, 8 and 10) when State Y was realized. The observed allocative efficiency (shown in numbers above the x-axis), as well as prices, are much closer to the predictions of the RE model than PI model. This experiment provided direct empirical evidence that markets can disseminate information from the informed to the uninformed through the process of trading alone, without an exchange of verbal communication. Such markets can achieve high levels of efficiency by transferring securities to the hands of those who value those most.

Evidence on the ability of markets to disseminate information led to a more ambitious experiment: Can markets behave as if diverse information in the hands of the traders be aggregated so it is in the hands of all? To address this question, Plott and Sunder [8] designed a market with three states of the world (X, Y, and Z). When the realized state was, say, X, they informed some traders that it was "not Y" and informed the others that it was "not Z," Do markets aggregate the diverse information in the hands of individual traders and behave as if everyone learns that the realized state is X in such a case? They found that in markets such aggregation and dissemination of diverse information can take place, and markets can achieve high levels of information and allocative efficiency. The same happens when investors have homogenous preferences (which make it easier for traders to infer information from the actions of others).

Just because markets can aggregate and disseminate information does not mean that all markets do so under all conditions. Experiments show that market conditions must allow investors the opportunity to learn information from what they can observe. Even in these simple experimental markets, these conditions are not always satisfied for various reasons (e.g., too many states, too few observations and repetitions to facilitate learning). For example, in the information aggregation experiment mentioned above, a complete market for three Arrow-Debreu securities is efficient, but an incomplete market for a single security is not.

Even in the best of circumstances, equilibrium outcomes are not achieved instantaneously. Markets tend toward efficiency, but cannot achieve it imme-

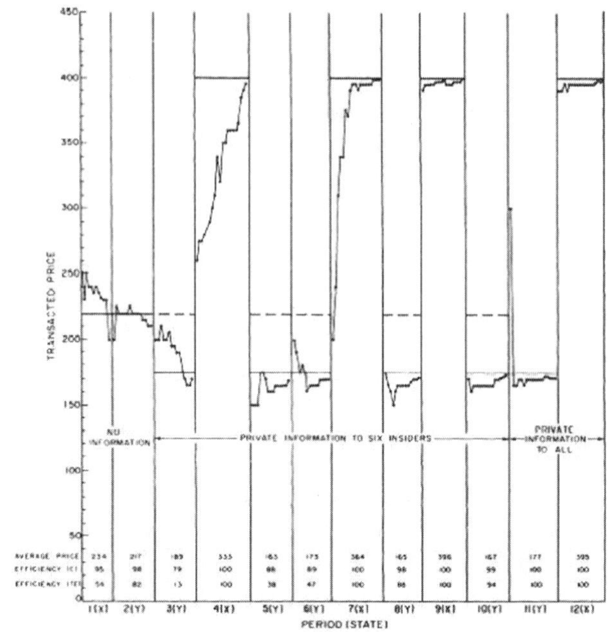

FIG. 4.—Market 3. Time series of contract prices

Fig. 1. Dissemination of Information in Security Markets (Source: Plott and Sunder, 1982, Figure 4)

diately. It takes time for investors to observe, form conjectures, test them, modify their strategies, etc. With repetition, investors get better at learning, but when the environment changes continually, including the behavior of other investors, the learning process may never reach a stationary point.

If markets are efficient in the sense of aggregating and disseminating information across traders, who would pay for costly research? Grossman and Stiglitz [3] and other authors have pointed out this problem. Experiments have helped us understand what actually goes on, and allowed us to better address this conundrum of the efficient market theory: Finite rate of learning makes it possible to support costly research, even in markets which tend toward efficient outcomes. Enough people would conduct research so the average returns to research equal the average cost. Research users have higher gross profits, but their net profits are the same as the profits of the others. As investors learn (in a fixed environment), their value of information decreases because they can ride free on others' information, and the market price of information drops. If the supply of information can be maintained at the lower price, the price drops to a level sustainable by learning frictions. If the supply of information also falls with its price, we get a noisy equilibrium.

After the exposure of misleading research distributed to clients from the research departments of investment bankers in recent years, regulators have sought to separate research and investment banking functions, and in some cases, required investment industry to fund free distribution of investment research to the public at large. The experimental research casts some light on the possible consequences of mandating the provision of free research to investors. It would be difficult, if not impossible, to assess the quality of such "free" research distributed to public. It is not clear if optimal investment in research can be maintained without private incentives to benefit from the results of the research. Mandated free distribution of research is likely to reduce its quality to a level where its price would be justified.

Economic theory tends to emphasize transaction prices as the main vehicle for the transmission of information in markets. Experimental markets show that other observables (bids, asks, volume, timing, etc.) also transmit information in markets. In deep markets, price can be the outcome of information transmitted through these other variables. In period 8 in Figure 1 for example, the first transaction occurs at the RE price. In order to arrive at the RE price, the traders need to learn information. This information transmission has already taken place through other variables before the first transaction of the period is executed.

Derivative markets help increase the efficiency of primary markets. Forsythe, Palfrey and Plott [1], in the first asset market experiment, showed that futures markets speed up convergence to equilibrium in the primary market (Friedman et al. [2]. Kluger and Wyatt [5] found that the option markets increase the informational efficiency of the equity market.

Traditionally, market efficiency has been defined statistically: if you can't make money from information (past data, public, or all information), the market is deemed to be efficient. Experiments have revealed that statistical efficiency is a necessary but not a sufficient condition for the informational efficiency of markets. The last four periods of the market depicted in Figure 2 from Plott and Sunder [8] are efficient by statistical criteria but are not informationally efficient. Just because you can't make money in the market does not mean that the price is right. Even when investors know that the price is not right, they may have no means of profiting from that knowledge.

5 What Is Next?

The above paragraphs give a highly selective summary of what we have learned from experimental asset markets. What is coming up next? The existence and causes of market bubbles is a perennial subject in financial economics. What might we learn about bubbles from experiments? Smith, Suchanek and Williams [9] showed that bubbles can arise in simple asset markets with inexperienced subjects, and tend to diminish with experience. Lei, Noussair and Plott [6] showed that bubbles can arise even when investors cannot engage in speculative trades. They suggest that bubbles can arise from errors in decision

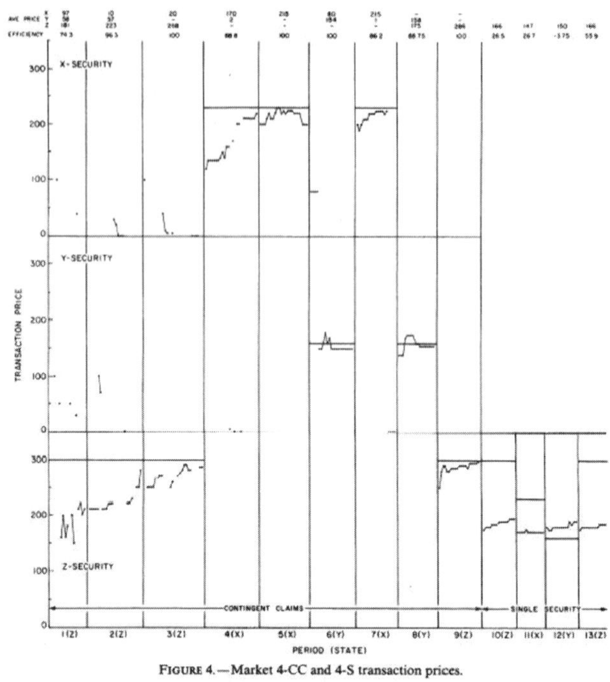

Fig. 2. Aggregation of Information in Security Markets (Source: Plott and Sunder [8], Figure 4)

making even in absence of a lack of common knowledge of rationality ("bigger fool" beliefs).

A recent experiment by Hirota and Sunder [4] explores the possibility that the fundamental economic model of valuation (DCF) may become difficult to apply in markets populated by short term traders. When a security matures beyond investment horizon, personal DCF includes the sale price at that horizon. The sale price depends on other investors' expectations of DCF beyond the investor's own horizon. Applying DCF involves backward induction from the maturity of the security through the expectations and valuations of the future "generations" of investors. Bubbles can arise, even with rational investors who make no errors, if they cannot backward induct the DCF. In their eleven experimental sessions, they found that bubbles arise consistently when the markets are populated with investors with short term investment horizons, and do not arise with long term investors.

DCF valuation model makes heroic assumptions about the knowledge necessary to do backward induction. Even if investors are rational and make no mistakes, it is unlikely that they can have the common knowledge necessary for the price to be equal to the fundamental valuation in a market populated by limited horizon investors. Not surprisingly, the pricing of new technology, high growth, and high risk equities are more susceptible to bubbles. In such circumstances, if we do not have common knowledge of higher order beliefs, testing theories of valuation becomes problematic.

This is only a thumbnail sketch of some experimental results on asset markets. We have not discussed many other important and interesting studies (for a more comprehensive survey, see Sunder 1995 in Kagel and Roth's Handbook of Experimental Economics). As the experimental camera focused on information processing in asset markets, the theoretical line drawing has been filled by details, shadows, color, and warts. This finer grain portrait of asset markets confirms the rough outline of the extant theory.But it is considerably more complex, and is providing guidance and challenges for further theoretical investigations of interplay of information in asset markets.

It is a unique experience to watch trading in an experimental asset market. You know all the information, parameters, and alternative theoretical predictions. Yet, what you see often surprises you, forcing you to rethink, and discover new insights into how these markets work. Experimental asset markets are our LEGO sets. Playing with them produces new ideas, and helps us sort out good ideas from the bad ones. I would have preferred to have you sit in my lab and experience all this yourself, instead of talking to you because students who participate in these markets gain a sophisticated understanding of the markets.

References

1. Forsythe, Robert, Thomas R. Palfrey, and Charles R. Plott. "Asset Valuation in an Experimental Market," *Econometrica* 50 (1982):537-68.
2. Friedman, Daniel, Glenn Harrison, and J. Salmon. 1983. "Informational Role of Futures Markets and Learning Behavior-Some Experimental Evidence," in *Futures Markets-Modelling, Managing and Monitoring Futures Trading*, M.E. Streit, ed. Oxford: Basil Blackwell.
3. Grossman, S. and J. Stiglitz. 1980. "On the Impossibility of Informationally Efficient Markets." *American Economic Review* 70:393-408.
4. Hirota, Shinichi, and Shyam Sunder, " Price Bubbles sans Dividend Anchors: Evidence from Laboratory Stock Markets," Revised October 2004.
5. Kluger, Brian D. and Steve B. Wyatt. 1990. "Options and Efficiency: Some Experimental Evidence," Working Paper, University of Cincinnati.
6. Lei, Vivian & Noussair, Charles N & Plott, Charles R, 2001. "Nonspeculative Bubbles in Experimental Asset Markets: Lack of Common Knowledge of Rationality vs. Actual Irrationality," *Econometrica*, 69(4), pages 831-59.
7. Plott, Charles R. and Shyam Sunder, "Efficiency of Controller Security Markets with Insider Information: An Application of Rational Expectation Models, " *Journal of Political Economy*, Vol. 90, No. 4 (August 1982), pp. 663-698.
8. Plott, Charles R. and Shyam Sunder, "Rational Expectations and the Aggregation of Diverse Information in Laboratory Security Markets," *Econometrica*, Vol. 56, No. 5 (September 1988), 1085-1118.

9. Smith, Vernon L., Gerry L. Suchanek, and Arlington W. Williams. 1988. "Bubbles, Crashes and Endogenous Expectations in Experimental Spot Asset Markets." *Econometrica* 56:6, 1119-1152.
10. Sunder, Shyam, "Market for Information: Experimental Evidence," *Econometrica*, Vol. 60, No. 3 (May 1992), 667-695.
11. Sunder, Shyam, "Experimental Asset Markets: A Survey," Chapter 6 in *Handbook of Experimental Economics*, A. Roth and John Kagel, eds., Princeton, NJ: Princeton University Press (1995), pp. 445-500.
12. Sunder, Shyam, "Real Phenomena, Theory and Design of Laboratory Experiments in Economics," July 2001.

Cheating in Markets: A Methodological Exploration *

Daniel Friedman

University of California, Santa Cruz

Summary. In the 1970s, experimental economics split from social psychology by embracing rational choice and equilibrium methods. Behavioral economics has recently narrowed the divide, to the dismay of some. The present paper argues that evolutionary dynamics provides a framework which unifies the best features of social psychology with equilibrium and rational choice.

Ongoing research in cheating in markets illustrates the main points. A new equilibrium model provides distinctive testable predictions under three regimes: autarky, frictionless free trade, and anonymous foreign trade with opportunities to cheat. The predictions organize quite well the data collected so far. Later phases of the project will allow trader networks to evolve, altering the market institution and perhaps affecting preferences. Thus the major forces recognized by social psychologists can be combined with a rationality and equilibrium to study how markets respond to the risk of cheating.

1 Introduction

The entire economics discipline has thrived on rational choice and equilibrium methods since the 1950s (e.g., Samuelson [30]). The subdiscipline of experimental economics took shape in the 1970s as pioneers showed how the same methods could guide laboratory experiments (e.g., Vernon Smith [32]; Charles Plott [28]). The researcher obtains clear and striking predictions from theoretical models that assume rational choice and equilibrium. Then the researcher recruits human subjects, induces appropriate preferences, information and economic institutions using clear and honest instructions, abstract

* The material on cheating in markets comes from ongoing research with Alessandra Cassar and Patricia Higino Schneider, funded by the US National Science Foundation under grant SES-0351801. Some of the methodological ideas were first presented at a November 1997 workshop at UCLA organized by Susanne Lohmann.

framing, and salient payments. Stationary repetition gives subjects the opportunity to adapt to the laboratory task, and outcomes observed after behavior settles down are compared to the theoretical predictions. Deviations suggest refinements of equilibrium theory, and the refinements in turn suggest further laboratory experiments. Thus one obtains a progressive research program organized by rational choice and equilibrium methods. (e.g., Davis and Holt [10]; Friedman and Cassar [12]).

Behavioral economics in recent years has followed a different script. The goal is to document departures from standard rational choice, especially irrationalities or other-regarding behavior, and to break the shackles of rational choice and equilibrium. Evidence comes mainly from laboratory experiments that often include salient pay and clear instructions, but that usually lack stationary repetition and that focus on the home-grown preferences of inexperienced subjects. Such experiments are as much in the style of social psychology as in the classic style of experimental economics.

The trend is quite dramatic. Many of the contributed papers in the present book are more in the new behavioral style than in the classical experimental economics style. The proportion of behavioral papers was considerably higher at the 2004 International Meeting of the Economic Science Association in Amsterdam, and was not much lower even at the 2004 North American regional ESA workshop.

Some experimental economists (young as well as old) think the pendulum has swung too far. It is all very well to broaden one's horizons, but something crucial is lost in abandoning traditional first principles. Experimental economics would be impoverished if it became a branch of experimental social psychology. At the same time, retreating to the classic approach of the 1970s does not seem the best way to advance experimental economics in the 21st century.

My suggestion is to use evolutionary dynamics as a framework that incorporates the best aspects of the classic experimental economics style as well as useful aspects of social psychology.

The next section summarizes the existing styles of economics and psychology and then sketches an evolutionary dynamics approach. To make the ideas concrete, the following sections introduce ongoing research regarding cheating and trust (central issues for social psychology) in markets (the home base of economic analysis). Section 3 poses the problem and summarizes new theoretical results in the classical style. Section 4 presents an experiment, again in the classical style, and the results obtained so far. Section 5 points to behavioral elements and offers an embracing framework in evolutionary dynamics. The concluding section offers some conjectures on other applications where the evolutionary dynamics approach might prove fruitful.

2 Three Methodologies

The classical economics methodology (e.g., Smith [32], who draws on Hurwicz [20] among many others) takes as given the characteristics of individual economic agents-e.g., preferences and endowments of resources, technology and information-and the economic institutions within which they interact, e.g., competitive markets. The researcher assumes that individual behavior is rational and has somehow achieved mutual consistency, e.g, competitive equilibrium (CE). The researcher then is able to deduce what the social outcomes (e.g., CE prices, quantities and gains from trade) must be, and to compare the observable components of the deduced outcomes to actual data obtained in the field or laboratory. Panel A of Figure 1 summarizes the classical economics methodology.

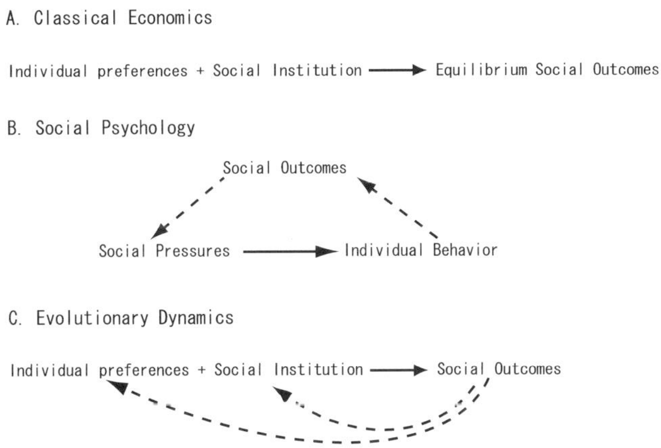

Fig. 1. Three Methodologies

The tradition in social psychology (e.g., Aronson [4]) is quite different. It is not so deductive, but the chain of reasoning starts with ideas about the pressures society exerts on individuals. These pressures shape individual preferences and behaviors, and reinforce social outcomes. The social outcomes are central, as shown in Panel B of Figure 1. As indicated by the dotted lines, social outcomes are the source of social pressures which typically shape individual behavior so as to perpetuate the social outcomes. Sometimes the pressures undermine themselves and lead to a changed society. Experiments are conducted mainly to clarify how social pressures affect individual behavior.

An example will illustrate the contrast between these two methodologies. Consider financial bubbles and crashes such as Tulipmania in 16th century Holland and England's South Sea Bubble a few years later, or Japan's stock market and real estate bubbles of the 1980s and California's dot.com bubble of the late 1990s. Historical accounts of these episodes often center on irrational "mob psychology." Indeed, Mackay [25], the classic treatment of the subject, is entitled Extraordinary Popular Delusions and the Madness of Crowds, and even the most popular treatment by an economist, Kindleberger [22], has a title that features psychological aberrations: Manias, Panics, and Crashes. These and many other books suggest that social pressures occasionally cause many people to do irrational things, and the irrationality can become contagious. Bubbles and crashes are individual irrationality writ large.

On the other hand, several authors in the last decade or so have explained some aspects of financial bubbles and crashes using the traditional economic methods of rational choice and equilibrium. Bikhchandani, Hirshleifer and Welch [6], and Banerjee [5] for example, show how equilibrium with fully rational agents can look something like a self-feeding irrational bubble. These and other authors demonstrate conditions under which it is rational for you to "follow the herd," even when your personal information suggests otherwise and you know that the herd sometimes thunders off in the wrong direction. A separate strand literature starting with Diamond and Dybvig [11] shows how "bank panics" can be explained without reference to social or clinical psychology. Under some well specified conditions, in equilibrium all depositors will try to withdraw their funds and the bank collapses. Morris and Shin [26] and a several later authors find conditions under which speculators can successfully attack a currency that is (in a reasonable sense of the word) fundamentally sound. Even the original Tulipmania (or much of the surviving evidence about it) can be rationalized using traditional economic methodology (Garber [15]).

The traditional economics approach has some impressive advantages. It starts from well-defined first principles, and employs flexible auxiliary assumptions (e.g., regarding the relevant economic institutions and information conditions) that often produce testable predictions. It is internally consistent, clear, and insightful. It can even provide economic insight into psychologically freighted words such as reputation, prestige, commitment and norms. There would be no reason for economists to consider other approaches if the predictions (or refinements of the predictions using sensible variations on the auxiliary assumptions) always enjoyed empirical success.

Unfortunately the world is not as tidy as one might wish, and empirical success is sometimes elusive. Garber's article illustrates the point that the combination of weak field evidence and flexible auxiliary assumptions makes it impossible to completely refute the traditional economics methodology. But laboratory studies can sharpen the empirical tests, and sometimes they undermine facile defenses of the traditional methodology. For example, consider the Ultimatum game introduced by Guth et al. [17]. After hundreds of theoretical and laboratory studies, it seems safe to say that the usual (subgame perfect Nash) equilibrium fails to predict outcomes very well, and that an explanation is needed for the fact that responders often reject small but positive offers.

How should economists proceed in the face of empirical failure? Do we have to give up our cherished traditional methods? Should we become merely specialized social psychologists?

I would urge economists not to follow psychologists in neglecting theoretical first principles. My reading (admittedly limited) of literature in cognitive and social psychology persuades me that it is difficult to maintain a coherent and broad based research program purely on the basis of empirical findings. A theoretical backbone helps keep researchers from wandering in circles (Lakatos [24]). But we need to take another look at the content of our first principles.

My first recommendation is to pay attention to the equilibration process. Traditional economics neglects the process by assuming that it automatically reaches completion (or nearly so) before outcomes are observed. Psychologists also neglect it by focusing on adjustments but seldom asking where they might settle down. We should ask ourselves what economic forces are pushing towards equilibrium and when (and whether) they might reach completion.

In general, I would say that economic adjustment processes operate on three distinct time scales. Individual learning is the most rapid. Given facilitating market and other social institutions, learning will rapidly and reliably produce outcomes close to equilibrium. Thus utility maximization and equilibrium should be thought of as the end result of a learning dynamic shaped by efficient social institutions. Unfortunately, social institutions evolve on a slower time scale and not always towards the efficient institutions. Our basic human nature as social creatures shapes the evolution of our institutions, and it in turn is subject to the slow force of genetic evolution. Thus the traditional economics approach will not always offer a reasonable approximation to actual outcomes.

To illustrate, consider again the Ultimatum game. Binmore [7] argues that for various reasons Responders learn much more slowly than Proposers, so the equilibration process converges not to the subgame perfect Nash equilibrium but rather to some particular imperfect Nash equilibrium in which Proposers usually make (and Responders usually accept) a substantial positive offer. An alternative explanation that I and several other economists favor involves non-standard preferences. Proposers often prefer the (0, 0) refusal outcome to an "unfair" proposed outcome like (8, 2); see Cox and Friedman [9] for a fully articulated model.

There is a real danger in departing from the standard view of preferences as exogenously fixed and selfish. All predictive power disappears when arbitrary behavior is rationalized by assuming arbitrary preferences for such behavior. To justify even the standard view of preferences, theorists traditionally have used evolutionary arguments, as exemplified in Alchian [1] and Friedman [14]. My point is that any other preference model requires the same justification. The model must account for the empirical data but also must pass the following theoretical test: people with the hypothesized preferences receive at least as much material payoff (or evolutionary fitness) as people with alternative preferences [1]. Otherwise, the hypothesized preferences would disappear over time, or would never appear in the first place. Papers such as Friedman and Singh [13] show how preferences for negative reciprocity (as in the previous paragraph) meet the theoretical test.

Panel C of Figure 1 summarizes the third social science methodology that I call evolutionary dynamics. Its first principle is evolution: preferences and institutions are not exogenous and arbitrary but rather are the products of well defined processes. At any moment of time, of course, preferences and social institutions are predetermined, as in the traditional economics methodology. They produce social outcomes, which are the traditional equilibrium outcomes when the institutions promote rapid learning that converges to those outcomes. For example, the Continuous Double Auction market institution promotes rapid convergence to Competitive Equilibrium in a variety of settings (e.g., Smith [32]; Cason and Friedman [9]).

The dotted lines in Panel C indicate slower feedback effects recognized by many psychologists but neglected in traditional economics. Individuals' emotional states (hence preferences over outcomes) respond quickly to their circumstances (e.g., to unfair treatment). Indirect evolution drives underlying emotional capacities and tendencies in the very long run. In the medium term (months to decades), cultural evolution shapes the economic institutions in which we interact. Thus traditional approach becomes a good short-to-medium term approximation when economic institutions promote rapid learning of the relevant equilibrium. When the traditional approach fails empirically, the methods of evolutionary dynamics can point to reasons for the discrepancy and offer a principled guide to better models.

The preceding discussion is rather abstract. Of course, the real test of a methodology is not how it sounds on first presentation, but rather how well it works in practice. The next several sections take work in progress to illustrate all three methodologies and to develop the abstract points just presented.

3 Cheating in Markets

Perhaps the most important theoretical result in traditional economics is that frictionless markets perform at 100% efficiency. Laboratory experiments since Smith [31] have given strong empirical support. Actual markets, however, face moderate to severe trading frictions. In particular, some buyers or sellers may cheat. The seller might ship an item of lower quality, and the buyer might not pay in full or on time. How do such markets perform?

The possibility of cheating seems to be a small friction in well-run modern markets, but it looms large in major markets of the ancient and medieval world

[1] This test is sometimes referred to as indirect evolution (Guth and Yaari [18]) because evolution operates on preference parameters that determine behavior rather than operating directly on behavior.

(e.g., Greif [16] and in some important contemporary markets. For example, according to some observers, eBay's main competitive advantage in Internet auctions is that its reputation rankings reduce such frictions (e.g., Anderson et al. [3]). Many observers believe that Russia's economic woes in the last decade are due largely to uncompetitive markets, in particular to weak enforcement of contracts (e.g., Klebnikov [23]). The observed volume of international trade is far smaller overall than predicted by traditional models, even when formal trade barriers are taken into account (Trefler [33]; Helliwell [19]). A leading suspect is the lack of trust in the enforcement of contracts in international markets (e.g., Rauch [29]; Anderson and Marcouiller [2]).

Cassar, Friedman and Schneider [8] develop a model in the classic economic tradition to study cheating in markets, as follows. Assume two distinct markets (called Red and Blue) with domestic supply and demand as in Figure 2. The textbook theory of Competitive Equilibrium (CE) predicts Autarky prices, transactions and surpluses as shown by the intersections of supply and demand in the first two Panels (Blue and Red). The same theory predicts the CE indicated in the third panel for frictionless free trade.

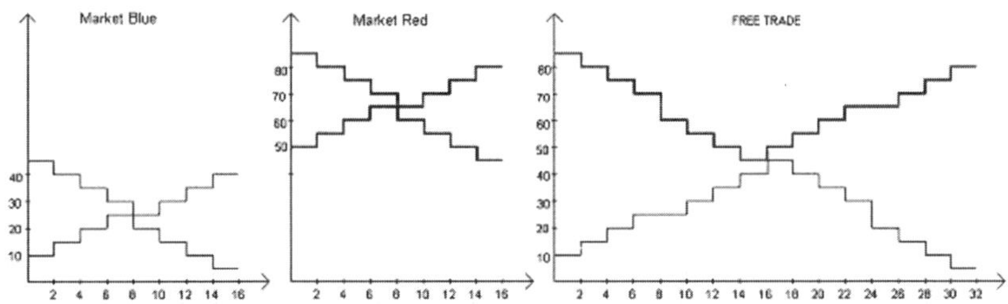

Fig. 2. Demand and Supply Schedules

The model of frictional trade gives the buyer in international transactions the option to pay only a given fraction π of the agreed price, and gives the seller the option to deliver only π of the value. Contracts are fully enforced in domestic transactions. Thus buyers and sellers must trade off better opportunities in international markets against the friction of imperfect contract enforcement.

It takes a little work, but the model can be solved explicitly to characterize CE with cheating for any value of π in [0,1]. At $\pi = 1$, of course, we have frictionless free trade, and at $\pi = 0$ the model predicts a reversion to Autarky. In between, the model yields novel testable predictions on price, volume and surplus. For example, agents with highest value and lowest cost are predicted to trade exclusively in their domestic market, while agents closer to the margin trade only in the cross market and always cheat; the overall volume is higher (!) than in frictionless free trade; and as π decreases from 1 to 0, domestic prices move non-linearly from autarky levels to the frictionless free trade level. Specific predictions are shown in Table 1.

Table 1. Testable Predictions

	Red market			Blue market			Cross-market					
							No Cheating			Cheating		
Autarky	65	8	160	25	8	160	-	-	-	-	-	-
Free Trade	-	0	0	-	0	0	45	16	640	-	-	-
Cheat Friction	60-65	6	150	30-35	6	150	-	-	-	40-45	10	125

4 An Experiment in the Classical Style

The traditional model just sketched provides the competitive equilibrium (CE) predictions. The next step in the classical tradition of experimental economics is to create a laboratory environment to test the predictions. The experiment reported in Cassar, Friedman and Schneider (2004) uses the well-known continuous double auction (CDA) market format for the reason mentioned earlier: it is known to promote rapid learning, at least in the Autarky treatment. At any instant during a CDA trading period, each buyer can post a public bid (offer to buy a unit at a given price or better) and each seller can post a public ask (offer to sell a unit at a given price or better). Each trader also at any instant can accept another trader's offer and immediately transact at the posted price p. A buyer with unit value v earns profit or surplus $v - p$ on the transaction, and a seller with cost c on the unit earns $p - c$, so the overall gains on the transaction are $v - c$.

The computerized CDA used in the experiment has several distinctive features, illustrated in Figure 3. Two markets, called Red and Blue, run simultaneously. Depending on the treatment, a trader may be able to trade only in her home market, or in both markets. Each trader can transact up to 4 units each period, and different units can have different cost or value. Each trader has an ID code that in some treatments can be used to identify her to potential transaction partners. The Figure shows the Marketplace window, active during the trading period. Between trading periods traders can view the History window by clicking the tab shown in Figure 1 above the upper edge of the open window. The History window shows all transactions from the period just completed, as well as a summary of trading profits from all previous periods.

Sixteen human subjects participate in each laboratory session. Four subjects are randomly assigned for the entire session to each of the four roles, buyer or seller in the Red or Blue home market. Each session with inexperienced subjects begins by going through the Autarky portion of the instructions, followed by a practice period and 3 to 4 Autarky trading periods.

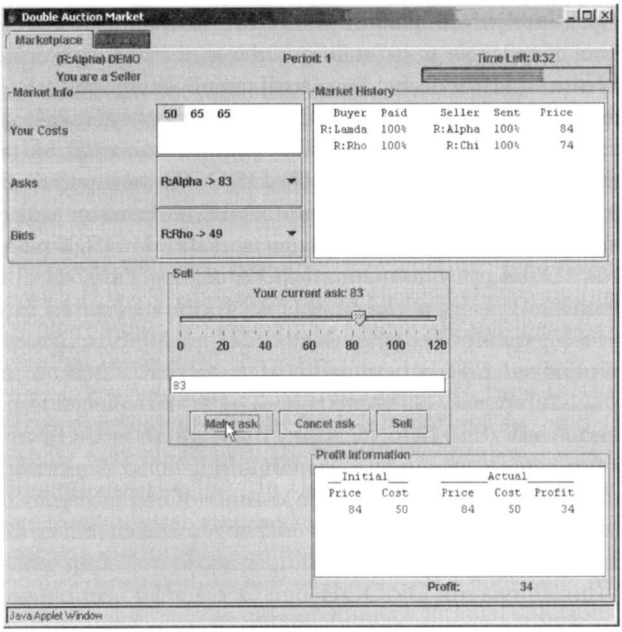

Fig. 3. User interface

`Autarky trading screen` is shown for a seller (ID code R:Alpha) in the Red market. In the center box she typed in an offer to sell a unit ('an ask') at price 83.
The upper left box labeled Market Info shows Alpha's costs for the current unit (highlighted, here 50) and remaining units (here 65 and 65). The next line shows the lowest ask (83, which here happens to be held by herself); clicking the small triangle pulls down a menu with other current asks. Bids (here the highest is 49 by R:Rho) similarly appear in the next line.
The Market History box (upper right) shows the current period transactions in the Red Market. Here, under autarky, all prices and trader ID's are shown.

Instructions, a practice period, and trading periods then follow for Frictionless Free Trade, and then for Cheating with $\pi = 0.5$. Sessions with experienced subjects skip Autarky and Frictionless Free periods, and reshuffle buyer values and seller costs once about half-way through the two-hour session. Each trading period lasts 240 seconds with a 20 second break between periods. After the last period, subjects are paid a \$5 show-up fee plus earnings for all periods; most subjects earn between \$15 and \$35.

Cheating is never allowed in domestic trades, e.g., in the Blue market between two Blue traders. The choices are sequential. First the trader accepting an offer chooses whether to cheat. That choice is observed by the trader who posted the offer, who then decides whether to cheat. Sellers cheat by delivering a good that costs πc instead of c and that provides value πv instead of v. Buyers cheat by paying only πp instead of p. Of course, the instructions avoid the word "cheat," and just talk about the choice of paying 50% (for π =0.5) or 100%, etc. It should be emphasized that in this treatment, cross market transactions are anonymous. Traders' ID codes are shown in all home market transactions, but are replaced by "??" when they post or accept bids and asks in the foreign market. The idea is to prevent traders from building reputations and to prevent discrimination among cross-market trading partners.

The results so far are very supportive of the competitive equilibrium model. Figure 4 summarizes the Autarky results. With inexperienced traders, actual prices converge towards CE from above in the low price Blue market, and converge from below in the high price Red market. In both markets, the standard deviation of prices declines and by the third period is less than 3.0, while the average price is within 2.0 of the CE predictions. Over all periods and in both markets, average price is within half a standard deviation of the CE prediction. Thus price convergence is quite sharp in both high and low price markets.

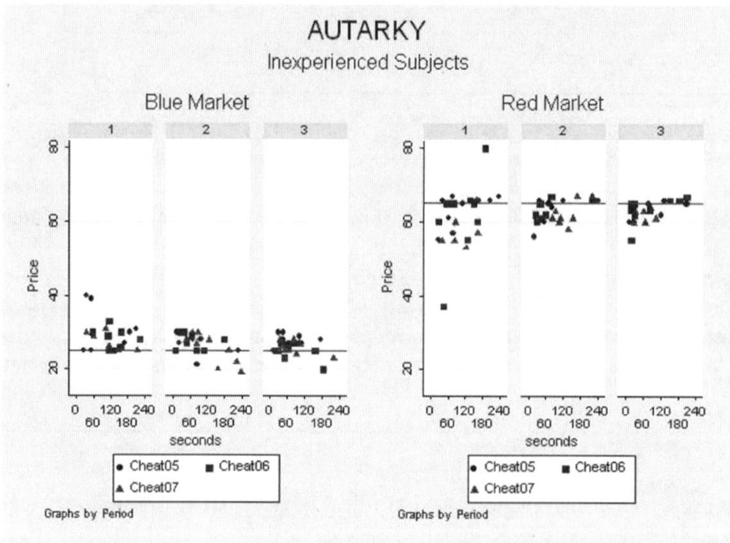

Fig. 4. AUTARKY (Inexperienced Subjects)

Average trading volume is within 1.0 unit of the CE prediction in every period in both markets, and the overall average volume of 8.1 is amazingly close to the 8.0 prediction. Average gains from trade are 259.4, about 81% of the CE prediction 320.0.

Figures 5 and 6 show behavior in frictionless free trade. As predicted, domestic trade volume shrivels, averaging 1 unit or less in both Red and

Blue. Average volume in the cross market reaches 15.7 in the third period and averages 15.0 overall, quite close to the CE prediction of 16.0. Average price in the cross market (and overall) are within 1 (or 2) of the CE prediction 45. However, the standard deviation of prices declines only slowly, to 7.2 in the third period versus 8.5 over all periods. CE surplus doubles to 640, and realized surplus also rises sharply to over 600. Indeed, overall efficiency is $604.4/640 \approx 94\%$.

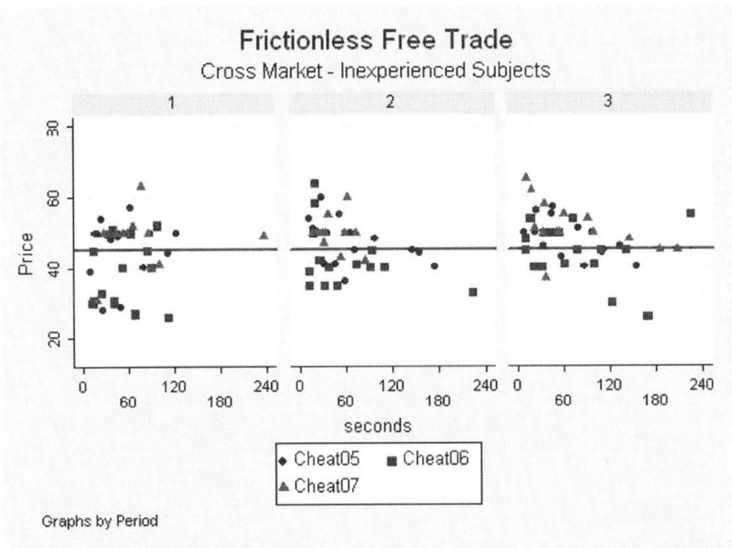

Fig. 5. Frictionless Free Trade (Cross Market – Inexperienced Subjects)

We conclude that actual behavior tracks the extreme CE predictions quite well in the first two treatments, but what happens with cheating frictions? Figures 7-10 and Table 2 show the results. With inexperienced subjects who have just finished the frictionless free trade treatment, the actual average number of cross-market trades with no cheating falls from 15.0 to 2.8, compared to the CE predicted fall from 16.0 to 0. Meanwhile, the average number of cross-market trades with cheating rises to 12.5, beyond the CE forecast of 10.0. With experienced traders, the average number of cross-market trades with no cheating falls to 1.5, and the average number with cheating falls to 11.0. Thus cheating is indeed rampant in cross-market trade (82% of trades for inexperienced and 88% for experienced traders), and deviations from the CE quantity predictions diminish with time and experience.

The predicted price range is 40-45 for cross market trades, and average prices are 46.9 for inexperienced and 48.3 experienced traders. The standard deviation of price declines over time, and averages about 7 for inexperienced

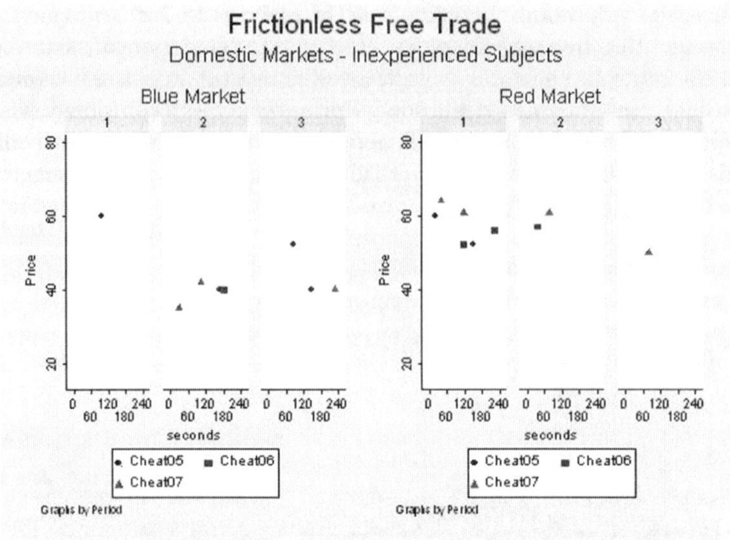

Fig. 6. Frictonless Free Trade (Domestic Markets – Inexperienced Subjects)

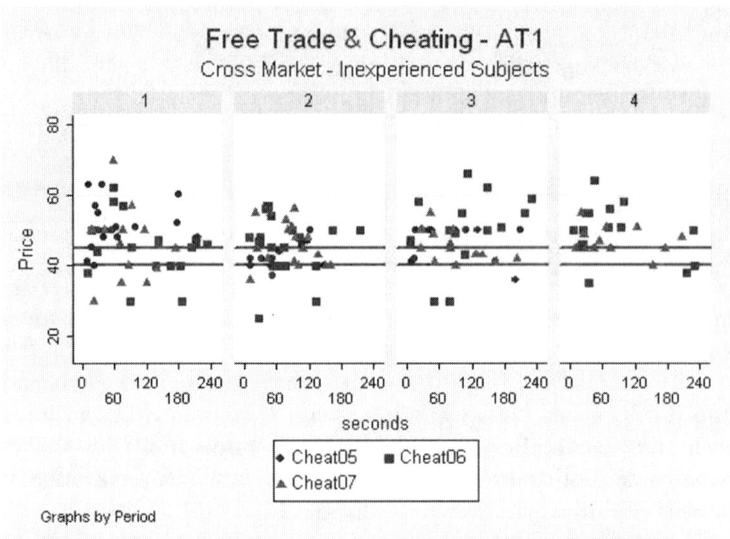

Fig. 7. Free Trade & Cheating - AT1 (Cross Market – Inexperienced Subjects)

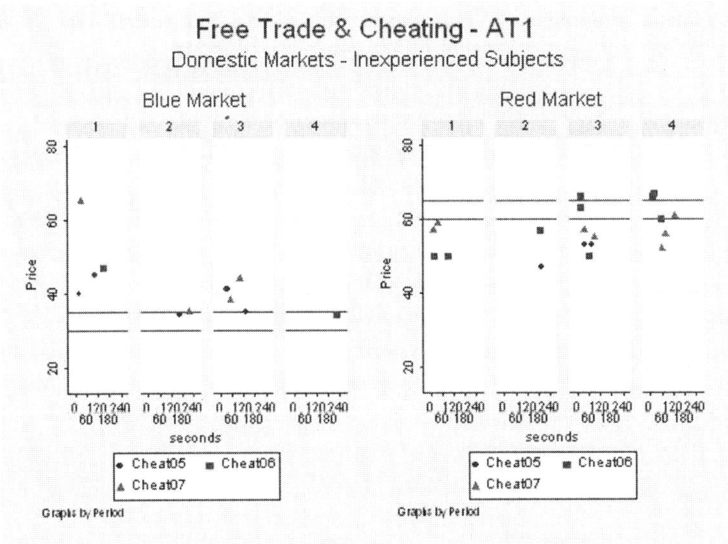

Fig. 8. Free Trade & Cheating - AT1 (Domestic Markets – Inexperienced Subjects)

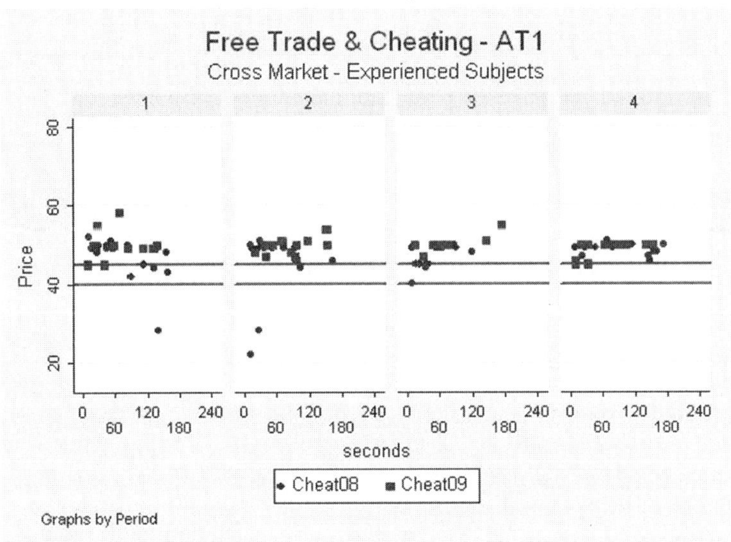

Fig. 9. Free Trade & Cheating - AT1 (Cross Market – Experienced Subjects)

Table 2. Mean Outcomes in Trade with Cheat Frictions (Inexperienced subjects). In parentheses: standard deviation and number of observations.

	Red Market			Blue Market		
Predicted	pR	qR	SR	pB	qB	SB
	60-65	6	150	30-35	6	150
Period 1	54.0 (4.7, 4)	1.3 (1.15, 3)	30.0 (30, 3)	49.3 (10.9, 4)	1.3 (0.6, 3)	15.0 (17.3, 3)
Period 2	52.0 (7.1, 2)	0.7 (0.6, 3)	16.7 (14.4, 3)	34.5 (0.7, 2)	0.7 (0.6, 3)	11.7 (12.6, 3)
Period 3	56.7 (5.8, 7)	2.3 (0.6, 3)	61.7 (20.2)	39.8 (3.4, 5)	1.7 (1.5, 3)	20.0 (18.0, 3)
Period 4	60.3 (5.7, 6)	3.0 (0, 2)	70.0 (21.2, 2)	34.0 (0, 1)	0.5 (0.7, 2)	2.5(3.5, 2)
Total 1-4	56.8 (6.0, 1.9)	1.7 (1.1, 11)	42.3 (29.3, 11)	41.6 (8.6, 12)	1.1 (0.9, 11)	13.2 (14.0, 11)

	Cross-market					
	NoCheating			Cheating		
Predicted	pNC	qNC	SNC	pc	qc	SC
	-	0	0	40-45	10	125
Period 1	46.0 (7.0, 15)	5.0 (2.6, 3)	238.3 (127.0, 3)	48.4 (9.7, 32)	10.7 (2.1, 3)	131 (21.9, 3)
Period 2	49.4 (5.0, 7)	2.3 (1.5, 3)	78.3 (53.5, 3)	44.6 (6.5, 44)	14.7 (2.1, 3)	261 (53.4, 3)
Period 3	50.0 (9.0, 3)	1.0 (1.0, 3)	53.5 (57.9, 3)	47.4 (7.3, 39)	13.0 (1, 3)	211 (40.3, 3)
Period 4	50.2 (7.4, 6)	3.0 (0, 2)	125.0 (21.2, 2)	48.4 (6.2, 22)	11.0 (2.8, 2)	151 (79.2, 2)
Total 1-4	48.0 (6.8, 31)	2.8 (2.13, 11)	123.6 (102.8, 11)	46.9 (7.6, 137)	12.5 (2.4, 11)	191.9 (67.9, 11)

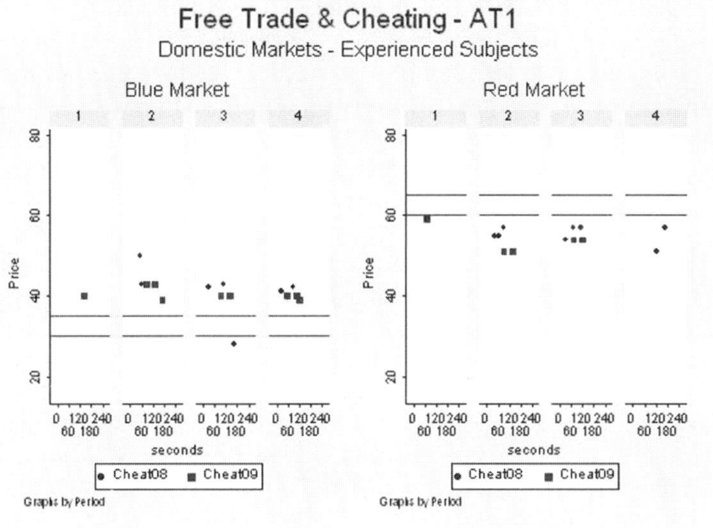

Fig. 10. Fred Trade & Cheating - AT1 (Domestic Markets – Experienced Subjects)

and less than 5 for experienced traders. Thus there is approximate price convergence to CE, a bit tighter than in the frictionless free trade treatment.

The CE model overpredicts trade volume in the two domestic markets. Actual volume in both markets under both treatments is less than half the predicted 6.0. Domestic price predictions are not bad: with inexperienced traders average actual prices fall within the predicted band by the last period, while with experienced traders the prices remain about 5 outside the predicted bands of 60-65 for Red and 30-35 for Blue.

Total surplus is quite variable in this treatment: 371 ± 71 with inexperienced traders and 403 ± 68 with experienced, versus the CE prediction of 425. The no-cheat cross-market trades increase surplus beyond the CE prediction, but the other departures from CE more than offset.

The CE predictions for this treatment again are rather extreme, and again behavior moves strongly in the predicted direction. We have less than complete convergence, however, and in particular the domestic trade volume does not recover to the predicted level.

5 Behavioral Explanations and Evolutionary Dynamics

Overall the traditional CE methods have led to surprisingly accurate predictions of behavior under the treatments examined so far: Autarky, Frictionless Free Trade, and Anonymous cheating. There clearly is a learning process at work. Many inexperienced traders at first do not cheat in cross-market transactions, but cheating seems contagious and becomes more prevalent in later periods. Or, to put it in less psychological terms, traders more frequently exercise their own cheat option after being cheated by another trader. This may be a dispassionate learning process, or may be done in anger, as part of negative reciprocity. Our current design does not distinguish between these two behavioral explanations. Either way, the data do indicate that the adaptation process proceeds rapidly, towards the competitive equilibrium in which cheating is universal in cross market transactions.

The next step in the research program involves institutional evolution. Traders will be able to break the anonymity constraint in cross market trading by becoming part of an interpersonal network. Long a staple of social psychology and sociology, personal networks are a crucial social institution that economists have only recently begun to study (e.g., Jackson and Watts [21]). After observing baseline efficiency and cheating in exogenously specified networks, my coauthors and I will run new treatments that allow traders to form their own personal networks. Thus there will be a feedback loop from social outcomes (prices, quantities and profits) to the institution, and we shall see what kinds of interpersonal networks emerge over time. We also expect to be able to see in more detail how individuals respond to cheating and to honest dealing both inside and outside their personal networks. Thus all links in the evolutionary dynamics scheme in Figure 1C will be operational.

I should try not to get too far ahead of the laboratory results here. Traditional equilibrium theory exists for pieces of the network trading environment. Perhaps these can be stitched together to obtain equilibrium predictions for which personal networks will form, and for individual behavior (and social outcomes) within those networks. Existing literature contains equilibrium models whose structure more or less resembles the trading networks we shall investigate. Some of the models feature inefficient equilibria, while others predict that efficient networks will emerge. We shall see what emerges in the lab.

6 Concluding Remarks

The evolutionary dynamics methodology outlined here might prove useful other ongoing research programs.

References

1. Alchian, Armen (1950) "Uncertainty, Evolution, and Economic Theory", *Journal of Political Economy, 58(3)*, 211-221.
2. Anderson, James A. and Marcoullier, Douglas (2000) "Insecurity and the Pattern of Trade: An Empirical Investigation", Revision of NBER Working Paper No. 7000.
3. Anderson, Steven, Daniel Friedman, Garrett Milam and Nirvikar Singh, "Seller Strategies on eBay", UCSC Working Paper #564, April 2004.

4. Aronson, Elliott (1984) *The Social Animal.* New York: Worth/Freeman.
5. Banerjee, Abhijit V., "A Simple Model of Herd Behavior", *The Quarterly Journal of Economics*, Vol. 107, No. 3. (Aug., 1992), pp. 797-817
6. Bikhchandani, Sushil; Hirshleifer, David and Welch, Ivo (1992) "A Theory of Fads, Fashion, Custom, and Cultural Change in Informational Cascades", *Journal of Political Economy, 100(5)*, 992-1026.
7. Binmore, Kenneth (2005). *Does Game Theory Work? Bargaining Experiments.* MIT Press.
8. Cassar, Alessandra; Friedman, Daniel and Schneider, Patricia (2004) "Cheating in Markets: A Laboratory Experiment", draft manuscript. `http://econ.ucsc.edu/~dan/`
9. Cason, Timothy and Friedman, Daniel (August 1996) "Price Formation in Double Auction Markets,", *Journal of Economic*, Dynamics and Control 20:8 1307-1337 .
10. Cox, James, Friedman, Daniel and Gjerstad, Steven (2004) "A Tractable Model of Reciprocity and Fairness", `http://leeps.ucsc.edu/leeps/projects/misc/tractable/tractable`
11. Davis, Douglas and Holt, Charles (1993), *Experimental Economics*, Princeton, NJ: Princeton University Press.
12. Diamond, Douglas W. and Philip H. Dybvig (1983) "Bank Runs, Deposit Insurance, and Liquidity", *The Journal of Political Economy*, Vol. 91, No. 3. pp. 401-419.
13. Friedman, Daniel and Cassar, Alessandra (2003) *Economics Lab: An Intensive Course in Experimental Economics.* New York: Routledge.
14. Friedman, Daniel and Singh, Nirvikar (2004) "Equilibrium Vengeance", `http://econ.ucsc.edu/~dan/`
15. Friedman, Milton (1953) "The Methodology of Positive Economics", in *Essays in Positive Economics.* Chicago, IL: University of Chicago Press.
16. Garber, Peter (1989) "Tulipmania", *Journal of Political Economy*, 97(3), 535-60.
17. Greif, A. (1993) "Contract Enforceability and Economic Institutions in Early Trade: The Maghribi Traders' Coalition", *American Economic Review*, 83(3), 525-548.
18. Guth, Werner; Schmittberger, Rolf and Schwarze, Bernd (1982), "An Experimental Analysis of Ultimatum Bargaining", *Journal of Economic Behavior and Organization*, 3(4), 367-88.
19. Guth, Werner and Yaari, Menachem (1992). "An Evolutionary Approach to Explaining Reciprocal Behavior", in U. Witt, ed., *Explaining Process and Change-Approaches to Evolutionary Economics.* Ann Arbor: The University of Michigan Press, 1992.
20. Helliwell, J.F. (1998) How Much Do National Borders Matter? Washington, D.C.: Brookings Institution.
21. Hurwicz, Leonid (1972) "On Informationally Decentralized Systems", in R. Radner and C. McGuire, eds., *Decision and Organization.* Amsterdam: North Holland.
22. Jackson, Matthew O.; Watts, Alison (2001) "The Existence of Pairwise Stable Networks", Seoul Journal of Economics, Vol.14, No.3, Fall, pp.299-321.
23. Kindleberger, Charles (1978) Manias, Panics, and Crashes, New York: Basic Books.

24. Klebnikov, Paul (2000) Godfather of the Kremlin: Boris Berezovsky and the Looting of Russia. New York: WW Norton & Co.
25. Lakatos, L. (1978) The Methodology of Scientific Research Programs, Volumes I and II. Cambridge: Cambridge University Press.
26. Mackay, Charles (1841) *Extraordinary Popular Delusions and the Madness of Crowds*,
27. Morris, Stephen and Shin, Hyun Song (1998) "Unique Equilibrium in a Model of Self-Fulfilling Currency Attacks", *American Economic Review*, 88(3), 587-597.
28. Plott, Charels (1982) "Industrial Organization Theory and Experimental Economics", *Journal of Economic Literature*, 20, 732-738.
29. Rauch, James (2001) "Business and social networks in international trade", *Journal of Economic Literature*, 39(4), 1177-1203.
30. Samuelson, Paul A. (1947) *Foundations of Economic Analysis*. Harvard University Press.
31. Smith, Vernon (1962) "An Experimental Study of Competitive Market Behavior", *Journal of Political Economy*, 70, 111-137.
32. Smith, Vernon (1982) "Microeconomic Systems as an Experimental Science", *American Economic Review*, 72, 923-955.
33. Trefler, Daniel (1995) "The Case of the Missing Trade and Other Mysteries", *American Economic Review*, 85(5), 1029-46.

Design Science: A Prelude *

Tatsuyoshi Saijo

Osaka University

Summary. Recently, we find the rise of criticisms in the field of mechanism design, which is to design a mechanism or system for achieving social goals such as efficiency and equity in the distribution of goods and services.

First of all, there is a criticism from experimentalists. In the verification of various mechanisms using human subjects in laboratories, these designed mechanisms do not necessarily function as prescribed. This fact itself is a criticism to not only this particular field, but also extending to economics as a whole. Departing from the initial stage of surprise with why theoretically expected results cannot be observed in laboratories, we are now entering the stage of determining why they do not function and what are the essential factors involved.

Second criticism involves the presumptions in the theories themselves. Mechanism design has not paid sufficient attention to information exchanges between people, the cost of processing, and the selections of equilibrium concepts. What is questioned now is the real validity of frameworks themselves, on which the theories are nested.

This report is, through exploratory works on issues of mechanism design, to contemplate hints of new approaches to the questions: what it means to design a mechanism; how to design them; and what shall be the next step economics needs to aim for.

1 Public Goods Provision

Let us consider the theory of public goods provision. When one person watches a TV program, it does not necessarily mean that that person excludes other persons from watching the same program. Such feature of public goods is called the non-rivalness of goods and services. Yet, TV programs can be scrambled to allow only those paying to watch them. In order words, those goods and services can exclude the possibility of consumption. Those goods and services that are non-rival but excludable are called public goods.

* Research was partially supported by the Grant in Aid for Scientific Research 15310023 of the Ministry of Education, Science and Culture in Japan.

In the textbook theory of public goods provision, there will be a free riding by the people who find it best to freely use public goods provided by others. Thus the level of public goods provision is short of a Pareto efficient level. Whether it was possible to design a mechanism to provide a Pareto efficient level of a public good or not was one of the main unsolved problems in 1970's.

Those forerunners who challenged this question were Clarke [7] and Groves [9]. They designed a mechanism, in which it would be best to express one's true preference of public goods regardless of other individuals' choices of strategies (i.e., satisfying strategy-proofness or incentive compatibility), although it would not be possible to attain Pareto efficiency. Their mechanism is mathematically equivalent to the second price auction in auction theory. Later, Green and Laffont [8] designed a mechanism that can not attain Pareto efficiency, but the allocation is very close to Pareto efficient one.

Those succeeded the initial studies were the groups of researchers called mechanism designers. Groves and Ledyard [10], Hurwicz [13], Walker [25], Varian [24] and others constructed games that would make Pareto efficient level of public goods achievable. In other words, they demonstrated that the Nash equilibrium allocations of such games is Pareto efficient. At about the same time, Maskin [18] provided necessary and sufficient conditions where the outcomes of a game coincide with a social choice correspondence, which provided theoretical background to mechanism designers.

By the time mechanism designers concluded that they could theoretically resolve the issue of public goods, Johansen [14] contradicted their approach itself. He pointed out that the framework of preference revelation would be far from the political process of public goods provision, and there was almost no incident when public goods were provided by preference revelation. He advocated for the analysis involving political process as the true analysis of public goods provision.

Later, many mechanism designers rejected Johansen's criticism and continued designing mechanisms, which were said to provide better performance. On the other hand, others started to question their approach from viewpoints different from Johansen's.

First of all, Kagel et al. [15], [16] verified in their experimental studies of the second price auction, which is strategy-proof, that people would rarely state their true valuation. People would usually state values higher than their true values. Moreover, Attiyeh et al. [2] and Kawagoe and Mori [17] confirmed in Clarke's pivotal mechanism experiments, which is mathematically equivalent to the second price auction, that the pivotal mechanism would not function either. They strongly questioned further study along such line.

Why strategy-proof mechanisms would not function in laboratories? Saijo, Sjöström, and Yamato [20] focused on the fact that most of strategy-proof mechanisms have a continuum of Nash equilibria, and considered that subjects participated in experiments might not necessarily choose dominant strategy even if that would state their true preference. In other words, which behavioral rules people would adopt would be entirely people's choice and not predetermined by researchers. They called the mechanism in which the outcomes of dominant strategy and Nash equilibria would agree as a secure mechanism and characterized it. They found that such a mechanism would hardly exist and only a special type of the Groves mechanism would be secure.

Moreover, Cason et al. [3] verified the performances of secure and non-secure mechanisms using subjects in the laboratory. They found that a secure mechanism did function and a non-secure mechanism did not.

The implication of aforementioned studies is important. It is because it concerns the very existence of a field of designing strategy-proof mechanisms, in which true preference announcement is a dominant strategy. Even if a strategy-proof mechanism can be designed successfully, however the mechanism itself may not function well, at least in laboratory, unless it is secure. Needless to say, such a mechanism is not likely to be applicable in a real society. If the mechanism is secure, still the possibility of its application is nil as long as it presumes the preference announcement as a strategy. How difficult it is for people to convey even a part of their preferences has been clearly demonstrated in the vote recounting event at Florida's Bush-Gore contention during the US Presidential election in 2000. Even if a preference can be represented by a continuous function, the dimensions of all possible preferences become infinite. It is principally impossible to exchange such information without costs.

Then, will the mechanism designed by, for example, Groves and Ledyard [10], which Nash-implements a Pareto efficient allocation, function well in a laboratory? Chen et al. [6] confirmed that the Groves-Ledyard mechanism does converge to Nash equilibria after the repetition of several hundred times with the same subjects. If a mechanism requires 100 repetitions to converge to an equilibrium, it will hardly have any practical use. In addition, Hamaguchi et al. [11] found through their emissions trading experiments that the Varian mechanism, that implement a social goal subgame perfectly, would not function well either.

Additional criticism concerning the mechanism design involves a tacit assumption concerning the public goods provision. Conventionally, mechanism designers assume tacitly that people should participate in the mechanism they designed. In other words, people must participate in it. What the non-excludability of public goods implies is that people will do free riding without participating in the mechanism. Saijo and Yamato [19] proved that, considering this factor, it would be impossible to design a mechanism in which every people participates. The issue of public goods provision has not been resolved in theories. Those challenging the issue of impossibility are young researchers such as Yu [26], Samejima [21], Shinohara [22] [23], Healey [12] and so on. Cason et al. [4] and Cason et al. [5] conducted experiments on this issue.

However, these criticisms have not provided solution to the question of how to design a system for providing public goods. This is because the circumstances assumed for models are far from the reality. Of course, the results of public goods provision model as idealtypus has significance, but to continue designing mechanisms by creating theoretical models in ignorance of how public goods have been provided in our society will be problematic.

In our society, public goods are not always improvided for or short supplied. For example, what we found during the 1980's and 1990's was rather the excess provision of public goods. Being aware of the fact that public goods could be short supplied if left alone, our ancestors tried on various methodologies to secure the provision of public goods, such as the "common land" to prevent the tragedy of the commons. It is certainly important to analyze these means, but have we conducted a thorough analysis on the decision making in public goods provision in this modern Japanese society?

Whether national level or community level, the decision making processes for public goods provision are normally structured as follows. First, bureaucrats are to prepare a draft of public goods provision policy. Then they announce the contents of the draft to relevant regional residents and hold public hearing meeting, in which mostly those people opposing the draft will likely participate and exercise strong influence over the revision of draft. Bureaucrats determine the strength and direction of the opposition opinions at the public hearing meeting, revise the draft, and resubmit to another public hearing meeting. Once this step is settled, bureaucrats submit the revised draft to the Council, which members are selected by bureaucrats and usually consisted of not only the experts of relevant public goods, but also prominent persons or stakeholders relevant to the introduction of the said public goods. Occasionally, some individuals of prominence or academic standing not residing in the region may become the Council members for the purpose of maintaining the neutrality. The Council reports to a Community leader such as a mayor or a governor, and the leader will acknowledge the result and move to implement the public goods provision project.

The study of such decision-making process has just begun in recent years. However, unless the public goods provision process undertaken today is fully analyzed, it is not possible to compare it with the mechanism proposed by mechanism designers. It is necessary to identify the pros and cons of each mechanism through comparison at least in theories. In order to adopt a mechanism that is theoretically more preferable than the current system, it is necessary to provide favorable results in laboratories, and in cases studies of other countries, other regions, or in the past. To sum up, mechanism designers incline to confine themselves in academic ivory tower and fail to propose mechanisms that can be alternatives to the existing systems.

2 Designing a Mechanism for Global Warming Mitigation

As the second example, let us consider the designing of a domestic system to prevent global warming under the UNFCCC's Kyoto Protocol. Once the

Kyoto Protocol enters into effect, Japan, for example, will have a GHG emissions cap at 94% of 1990 emissions for 5 years from 2008 to 2012. If the actual emissions do not exceed this cap, Japan will have the options to either bank the difference between the cap and actual emissions to be used after 2013, or to sell them to other countries. If the actual emissions exceed the cap, on the other hand, it must purchase emissions reduction from other countries.

To achieve such targets, there are various potentially conceivable systems. Discussed below is the review of several approach designs so to identify economically correct approaches and to contemplate on what should be done to the new field of design science. In the designing of systems discussed below, I will try to approach problems through exaggerated profiling of systems' characterization, rather than through minute examination of details.

Upon designing systems, various indexes can be applied for different approaches of assessment. Economists may stress "economic efficiency" to minimize GHG reduction costs. On the other hand, from the viewpoint of absolute compliance with Kyoto targets, the first priority will undoubtedly be the "compliance" of the Kyoto targets. From the viewpoint of attaining as much GHG reduction as possible, rather than mere compliance with the Kyoto targets, then the system must aim for global "environmental" conservation.

There is no single incentive to encourage people to achieve such targets. Those in emissions trading business will certainly recommend emissions trading, without even considering the "economic efficiency." Bureaucrats involved in policy-making to promote energy saving technologies or vested interests at the back of such policy will advocate "environmental conservation" in order to secure budgets for such policy, rather than "economic efficiency" or "compliance." I have no intention to discuss "good or bad" of incentives. Important point here is that researchers responsible in creating "design science" have not contemplated fully on past incentives as demonstrated below.

Researchers of expertise in strategy-proofness will think of a game to state GHG reduction technology without questioning. This means that each entity is to state a reduction technology function, but it is easy to show that the true technology function announcement will not be the best way. Moreover, if the mechanism is a Clarke type though sacrificing efficiency, they will show that the true technology function announcement is the dominant strategy. As discussed in the previous section, researchers in this field will not likely analyze who shall collect information by what methods, how such information is processed and in what way the distribution of reductions to each country can be determined. Those involved in policy-making will likely consider such proposal as a thing in the air.

Mechanism designers will undoubtedly demonstrate that they can design a mechanism to achieve efficiency. For instance, the Varian mechanism will enable the achievement of Kyoto target through sub-game perfect equilibrium. As in the case of strategy proofness, however, problem of information processing will not be addressed. There is much unnaturalness embraced in mechanisms designed by mechanism designers such as the Varian mechanism.

For example, in a numerical statement game, if all but one state the same numerical value, then the agent not stating that value will have penalties. Sometimes, there may be a designer who will design a mechanism to enable the confiscation of all the assets of this agent. If one loses all the assets by failed speculation in a futures market, then one has only oneself to blame. However, why is it that one merely providing information different from others has to receive penalty? What is the legal basis to put penalty to such an entity? Still, mechanism designers continue designing mechanisms without questioning the "unnaturalness" of their mechanisms. Empirical researchers, on the other hand, verify that such mechanisms will not function as prescribed. It is deemed that policy-makers will not seriously consider such mechanisms as an alternative.

Of course, the well known economic tools in the field are carbon tax and emissions trading. According to the standard textbooks, both approaches are said to bring the efficient compliance of the Kyoto target. Let us first examine the carbon tax, which many researchers recommend. Actually, no one knows what can be the rate of carbon taxes that can achieve how much of the target. Moreover, even if carbon tax is imposed, its rate cannot be changed easily, since such changes need the approval of the Diet. The fact that laboratory experiment has not yielded any proper rate of carbon tax to enable Kyoto target achievement, as explained in Akai, Kusakawa et al. [1], indicates the difficulty of complying with the Kyoto target through carbon tax. In other words, carbon tax, though efficient, is not fit to achieve a pre-fixed target. As shown here, a proposal not contemplating on political restrictions will not be justifiable.

How about emissions trading? According to the standard textbooks, emissions trading also enables efficient achievement of a fixed target. In other words, it can provide both efficiency and compliance. However, according to the empirical study by Akai, Kusakawa, et al. [1], there can be non-textbook cases if any uncertainties of emissions reduction investments.

3 Framework of Design Science

How one needs to design a system with no precedents, such as the case of domestic system design for global warming? No well functioned mechanism can be designed, if only relying on the approaches in a specific field. What is needed is to design various alternatives by setting multiple number of assessment criteria, and using multiple approaches. As discussed in the previous section, the mechanisms designed by current mechanism designers are not likely to become alternatives.

When developing various alternatives for domestic mechanism design to prevent global warming, one must consider designing a system that can incorporate characteristics unique to Japan and non-existent in other countries (such as almost total reliance on imports of fossil fuels, difficultly to comply with the Kyoto target, and so on). Instead of relying on a sole approach, we must consider the combination of multiple methodologies. Also, it is necessary to develop new methodologies, rather than solely relying on the conventional ones. This is where researchers can exhibit their originalities. If one alternative is inferior to other alternatives in the light of every assessment criteria, then that cannot be an alternative. Thus, what would remain are only those alternatives that cannot totally surpass other alternatives. Such an attempt was made by Akai, Okagawa, Kusakawa, et al. (2004). However, there have been almost no studies that squarely address the issue of incentives for various stakeholders in the process of policy decision making. Amazingly, economists including environmental economists hardly ever study the comparison of systems, and merely propose a system they consider preferable, or introduce systems of other countries.

If multiple alternatives remain as theoretical proposals, then they cannot become truly adoptable proposals. Verification is needed to determine whether each proposal can exhibit theoretically prescribed performance or not by using various methodologies. Also needed to implement is the verification of measurement models and numerical calculation models such as applied general equilibrium model, proof of each system in laboratories using subjects, and confirmation of a system or similar system of the past or in other countries, through the survey of their successes and failures.

After these processes, the proof of each proposal's performance should be implemented, and any problems likely to be arisen should be solved. Then, coming would be the works to submit and to verify new proposal or revisions of existing proposals.

Even if a good proposal is made, it can be wasted unless the Council takes it up as its agendum. If the proposal is not compatible with incentives for stakeholders or bureaucrats at the back of the Council, then the proposal will not be accepted. Also important is the process of letting policy-makers understands various proposals. For instance, not many people understand the "marginal concept," which is a common knowledge for economist, or comprehend the meaning of "economic efficiency" and "dead weight loss."

Preferable system will be to establish a center to design systems and policies independent of bureaucrats, rather than the system for bureaucrats to expend national budget to a Council set for each issue and to consign studies to think tanks and universities of the private sector. Policy makers will be exposed to competition with systems and policies recommended by external research institutes. Of course, the Council style policy-making process itself will become an important research subject for such centers.

References

1. Akai, K., A. Okagawa, T. Kusakawa and T. Saijo (2004): "Designing Domestic Institutions to Cope with Global Warming," mimeo (in Japanese).
2. Attiyeh, G., Franciosi, R. and R. M. Isaac (2000): "Experiments with the Pivotal Process for Providing Public Goods," *Public Choice*, 102, 95-114.
3. Cason, T., T. Saijo, T. Sjöström and T. Yamato (2004): "Secure Implementation Experiments: Do Strategy-proof Mechanisms Really Work?" mimeo.
4. Cason, T., T. Saijo, and T. Yamato (2002): "Voluntary Participation and Spite in Public Good Provision Experiments: An International Comparison," *Experimental Economics*, Vol. 5, pp.133-153.
5. Cason, T., T. Saijo, T. Yamato and K. Yokotani (2004): "Non-Excludable Public Good Experiments," *Games and Economic Behavior*, 49 pp.81-102.
6. Chen, Y. and R. Gazzale (2004): "When Does Learning in Games Generate Convergence to Nash Equilibria? The Role of Supermodularity in an Experimental Setting," *American Economic Review*, forthcoming.
7. Clarke, E. H. (1971): "Multipart Pricing of Public Goods," *Public Choice*, 2, 19-33.
8. Green, J. R. and J.-J. Laffont (1979): *Incentives in Public Decision Making*, Amsterdam: North-Holland.
9. Groves, T. (1973): "Incentives in Teams," *Econometrica*, 41, 617-31.
10. Groves, T., and J. O. Ledyard (1977): "Optimal Allocation of Public Goods: A Solution to the 'Free-Rider' Problem," *Econometrica*, 45, 783-809.
11. Hamaguchi, Y., T. Saijo, and S. Mitani (2003): "Does the Varian Mechanism Work?: Emissions Trading as an Example" *International Journal of Business and Economics*, Vol. 2, No. 2, pp.85-96.
12. Healy, P.J. (2004): "Equilibrium Participation in Public Goods Allocations," mimeo.
13. Hurwicz, L. (1979): "Outcome Functions Yielding Walrasian and Lindahl Allocations at Nash Equilibrium Points," *Review of Economic Studies*, 46, 217-225.
14. Johansen, L. (1977): "The Theory of Public Goods: Misplaced Emphasis?," *Journal of Public Economics*, Vol.7, 147-152.
15. Kagel, J. H., R. M. Harstad and D. Levin (1987): "Information Impact and Allocation Rules in Auctions with Affiliated Private Values: A Laboratory Study," *Econometrica*, 55, 1275-1304.
16. Kagel, J. H. and D. Levin (1993): "Independent Private Value Auctions: Bidder Behavior in First- Second- and Third-Price Auctions with Varying Number of Bidders," *Economic Journal*, 103, 868-879.
17. Kawagoe, T. and T. Mori (2001): "Can the Pivotal Mechanism Induce Truth-Telling? An Experimental Study," *Public Choice*, 108, 331-354.
18. Maskin, E. (1977): "Nash Equilibrium and Welfare Optimality," appeared in *Review of Economic Studies*, 66,23-38
19. Saijo, T., and T. Yamato (1999): "A Voluntary Participation Game with a Non-Excludable Public Good," *Journal of Economic Theory*, Vol.84, pp.227-242.
20. Saijo, T., T. Sjöström and T. Yamato (2004): "Secure Implementation," mimeo.
21. Samejima, Y. (2003): "Inducing Participation in Lindahl Mechanisms," mimeo.
22. Shinohara, R. (2003): "Coalition-proof Equilibria in a Voluntary Participation Game," mimeo.
23. Shinohara R. (2004): "Strong Equilibrium in a Participation Game with a Discrete Public Good", mimeo.

24. Varian, H. R. (1994): "A Solution to the Problem of Externalities When Agents Are Well-Informed," *American Economic Review*, 84, 1278-1293.
25. Walker, M. (1980): "On the Nonexistence of a Dominant Strategy Mechanism for Making Optimal Public Decisions," *Econometrica*, 48, 1521-1540.
26. Yu, Z. (2001): "A Strategic Trade and Environmental Policy: Argument for the Kyoto Protocol," mimeo.

Part III

Non Game-Theoretic Individual Decision Making

Separation of Intertemporal Substitution and Time Preference Rate From Risk Aversion: Experimental Analysis With Reward Designs*

Ryoko Wada[1] and Sobei H. Oda[2]

[1] Keiai University
[2] Kyoto Sangyo University

1 Introduction

In the standard intertemporal specification of expected utility

$$U_t = (1-\beta)E\left[\sum_{t=0}^{\infty} \frac{c_t^{1-\sigma}}{1-\sigma}\beta^t\right], \tag{1}$$

the coefficient of relative risk aversion σ is the reciprocal of the rate of intertemporal substitution $1/\sigma$. This has been suspected as a source of poor performance of the standard stochastic consumption model and the risk-premium puzzle in asset pricing [1]. More generally, the problematic feature of expected utility applied to intertemporal settings is that its treatment of 'gambling over time' cannot distinguish risk aversion and intertemporal substitution [2].

Epstein-Zin introduced a recursive model of non-expected utility in which risk aversion is separated from intertemporal substitution. It has the form

$$U_t = \left[(1-\beta)c_t^{\rho} + \beta\left(E_t\widetilde{U}_{t+1}^{\alpha}\right)^{\rho/\alpha}\right]^{1/\rho}, \tag{2}$$

where α explains risk aversion applied only to 'timeless gambles' and ρ explains intertemporal substitution. Thus attitudes toward gambling over time are clearly decomposed. The model generalizes the standard one in the sense that expected utility hypothesis is maintained only for timeless gambles. When $\alpha = \rho$, model (2) reduces to model (1).

Our objective is to examine the above noted separation as well as the validity of recursive utility, by means of experimental methods.Econometric

* The author wishes to thank Dr. Hayashi Takashi and Dr. Stahl Dale O. at Taxes University for their precious advices.

[1] See Hall [3], Mehra and Prescott [5].

[2] See Kreps and Porteus [4] for more theoretical arguments.

estimation approaches have been taken by Epstein and Zin [2], using time-series data of aggregate consumptions and asset returns. This study show a troubling pattern that estimation exhibits negative rate of time preference. We adopt experimental methods to make the setting more controllable.

2 Questions and Procedures to Test Recursive Utility

The procedure of our test of recursive utility is divided into four steps. 1. Estimation of $(\frac{1-\beta}{\beta})^{\frac{1}{\rho}}$ (Q1–Q3) 2.Estimation of β and ρ (Q4) 3.Estimation of α (Q5,6,8) 4.Test of dynamically consistent choice (Q7,9)

2.1 Estimation of the Range of $(\frac{1-\beta}{\beta})^{\frac{1}{\rho}}$

The first step is to estimate the range of $(\frac{1-\beta}{\beta})^{\frac{1}{\rho}}$. Q1 is the most basic question.

Q1. WHICH DO YOU PREFER, A(10,0) OR B(0,10) ?
A(10,0) MEANS THAT YOU GET 1000 YEN TOMORROW AND NOTHING 29TH DAYS AFTER. THE UNITS OF THE NUMBER IS 100 YEN. IF YOUR ANSWER IS A(B), PLEASE ANSWER Q2(3).

Table 1. Procedure of Q2 and 3

Base choice	Q2(Q3)-1	Q2(Q3)-2	Q2(Q3)-3	Q2(Q3)-4	Q2(Q3)-5	Q2(Q3)-6
Q2 **A**(10, 0)	$C(0, 10.1)$	$D(0, 10.5)$	$E(0, 11)$	$F(0, 12)$	$G(0, 15)$	$H(0, 20)$
Q3 **B**(0, 10)	$C_A(10.1, 0)$	$D_A(10.5)$	$E_A(11, 0)$	$F_A(12, 0)$	$G_A(15, 0)$	$H_A(20, 0)$

If a subject answers A in Q1, the $(\frac{1-\beta}{\beta})^{\frac{1}{\rho}}$ of him/her is greater than or equal to 1. If a subject answers B in Q1, $(\frac{1-\beta}{\beta})^{\frac{1}{\rho}}$ is smaller than or equal to 1. We exclude the answer that the subject is indifferent between the alternatives,to make it easier for subject to respond. From the answers to Q2 we infer how close to 1 the subject's $(\frac{1-\beta}{\beta})^{\frac{1}{\rho}}$ is. See Table 1 for their procedure. Pattern of the answers and the corresponding values of $(\frac{1-\beta}{\beta})^{\frac{1}{\rho}}$ are as in Tables 2.

For illustration, suppose that the subject chooses A in Q1 and her choices in Q2 reveal that she ranks $A(10,0)$ between $E(0,11)$ and $F(0,12)$. In the model, the value of deterministic consumption stream is given by

$$U = [(1-\beta)c^{\rho} + \beta z^{\rho}]^{\frac{1}{\rho}}$$

where c denotes the current consumption and z denotes the future consumption. Hence the value of $A(10,0)$ is $10(1-\beta)^{\frac{1}{\rho}}$ and the values of $E(0,11)$, $F(0,12)$ are $11\beta^{\frac{1}{\rho}}$, $12\beta^{\frac{1}{\rho}}$, respectively. Thus we have

$$11\beta^{\frac{1}{\rho}} \leq 10(1-\beta)^{\frac{1}{\rho}} \leq 12\beta^{\frac{1}{\rho}},$$

which implies

$$1.1 \leq \left(\frac{1-\beta}{\beta}\right)^{\frac{1}{\rho}} \leq 1.2.$$

Without loss, we approximate $\left(\frac{1-\beta}{\beta}\right)^{\frac{1}{\rho}}$ by the midpoint of the range, 1.15.

Table 2. Estimation procedure for $\left(\frac{1-\beta}{\beta}\right)^{\frac{1}{\rho}}$

pattern	Q2-1	Q2-2	Q2-3	Q2-4	Q2-5	Q2-6	Range of $\left(\frac{1-\beta}{\beta}\right)^{\frac{1}{\rho}}$	Estimate
1	A	A	A	A	A	A	$[2, -]$	2
2	A	A	A	A	A	H	$[1.5, 2]$	1.75
3	A	A	A	A	G	H	$[1.2, 1.5]$	1.35
4	A	A	A	F	G	H	$[1.1, 1.2]$	1.15
5	A	A	E	F	G	H	$[1.05, 1.1]$	1.075
6	A	D	E	F	G	H	$[1.01, 1.05]$	1.03
7	C	D	E	F	G	H	$[1, 1.01]$	1.005

Table 3. Procedure of Q4

Q4-1	Q4-2	Q4-3	Q4-4	Q4-5	Q4-6	Q4-7	Q4-8
$U1(4.9, 4.9)$	$U2(4.7, 4.7)$	$U3(4.5, 4.5)$	$U4(4, 4)$	$U5(3.5, 3.5)$	$U6(3, 3)$	$U7(2.5, 2.5)$	$U8(2, 2)$

2.2 Estimation of β and ρ

The values of β and ρ of a subject is determined by the previous estimation of her $\left(\frac{1-\beta}{\beta}\right)^{\frac{1}{\rho}}$ and her answers to Q4.

For illustration, suppose that the previous choices made by the subject determines her $\left(\frac{1-\beta}{\beta}\right)^{\frac{1}{\rho}}$ to be 1.15, and her choices in Q4-1 to Q5-8 reveal that she ranks $A(10,0)$ between $U3(4.5,4.5)$ and $U4(4,4)$. Since $U3(4.5,4.5)$ and $U4(4,4)$ give constant streams, their values of U are 4.5, 4, respectively. Thus we have

$$0.4 \leq (1-\beta)^{\frac{1}{\rho}} \leq 0.45.$$

We approximate the value $(1-\beta)^{\frac{1}{\rho}}$ of 0.425 by the midpoint of the range. By combining this with a system of equations,$(\frac{1-\beta}{\beta})^{\frac{1}{\rho}} = 1.15$ which leads to $\beta = 0.473$ and $\rho = 0.75$.

2.3 Estimation of Risk Aversion for Static Gambles

Next, we estimate risk aversion. For completeness we start with estimating risk aversion for static gambles.

Q5 It is decided to get money of 1 thousand yen tomorrow. Which do you prefer regarding money of the 29th day after?

X: A certain 1 thousands yen.
Y: A lottery ticket: you get 2 thousand yen if you win, and you get nothing if you lose.

If a subject chooses X in Q5, he is required to answer from Q6 and Q7. The subjects who answer Y are risk neutrals or risk lovers. We estimate the strength of the subject to avoid risk in static gambles in Q6. The range of α is estimated by the pattern of answers of Q6.

Table 4. Procedure of Q6

Base choice	Q6-1	Q6-2	Q6-3	Q6-4	Q6-5	Q6-6	Q6-7
$Y\begin{cases}20\,;0.5\\0\,;0.5\end{cases}$	X_{F1} Certain 9	X_{F2} Certain 8	X_{F3} Certain 7	X_{F4} Certain 6	X_{F5} Certain 5	X_{F6} Certain 4	X_{F7} Certain 3

2.4 Procedure for Estimating "Intertemporal" α

Next, we estimate the degree of intertemporal risk aversion using Q7-1-Q7-7.

Table 5. Procedure of Q7

Base choice	Q7-1	Q7-2	Q7-3	Q7-4	Q7-5	Q7-6	Q7-7
$Y\begin{cases}20\,;0.5\\0\,;0.5\end{cases}$	$X1$ Certain 10	$X2$ Certain 6	$X3$ Certain 5	$X4$ Certain 4	$X5$ Certain 3	$X6$ Certain 2	$X7$ Certain 1
Next year	This year	This year	This year	This year	This year	This year	This year

We describe the procedure to estimate the values of intertemporal α. For illustration, suppose her choices in Q9 reveal that she ranks the lottery $Y(20; 0.5, 0; 0.5)$ given next year between $X3$ = Certain 5 this year and $X4$ = Certain 4 this year. Recall in equation (2), z denotes the future consumption which is risky. Then the value of $Y(20; 0.5, 0; 0.5)$ given next year is $\left[\beta\left(0.5 \cdot 20^{\alpha}\right)^{\rho/\alpha}\right]^{1/\rho}$ whereas the values of $X3$ = Certain 5 this year and $X4$ = Certain 4 this year are $5(1-\beta)^{\frac{1}{\rho}}$ and $4(1-\beta)^{\frac{1}{\rho}}$, respectively. Thus we have

$$4(1-\beta)^{\frac{1}{\rho}} \leq \left[\beta\left(0.5 \cdot 20^{\alpha}\right)^{\rho/\alpha}\right]^{1/\rho} \leq 5(1-\beta)^{\frac{1}{\rho}},$$

which implies

$$0.2\left(\frac{1-\beta}{\beta}\right)^{\frac{1}{\rho}} \leq (0.5)^{\frac{1}{\alpha}} \leq 0.25\left(\frac{1-\beta}{\beta}\right)^{\frac{1}{\rho}}.$$

Since β and ρ are known by the previous steps, we estimate the range of α. The tests regarding to recursive utility are two-folds. First, we test whether the choices made by the subjects are dynamically consistent, that is, whether the subjects fall in the model of recursive utility. We judge a subject to be time consistent if there is some range of α to fulfill both ranges for static gamble and for dynamic gamble.

Second we test the separation of risk aversion from intertemporal substitution within the set of subjects who fall in recursive utility.

3 Results and Conclusions

We compare the difference of results of experiments in table 6- 10.

1. We onformed that each participant's decision-making depends upon three factors: the time preference rate, preference for smoothness of intertemporal receipts, and the narrowly defined risk attitude.
2. Reasonable part of the subjects exhibit dynamically consistent choices. Also, as the time-consistent subjects, we observed their intertemporal decision making is not explained by the expected utility hypothesis.
3. Our results of positive time preference rates contrasts with the result in Epstein-Zin [2]. This suggests that experimental methods have certain advantage over estimation approaches.

Table 6. The comparison with the previous experiment: designs and results

experimental design	*present experiment*	*Wada and Oda [6](WO)*
executiontime	April - June 2005	October, November 2004
number of subjects	39 at Keiai	25 at Keiai, 89 at KyoSan
reward designs	with reward designs	without reward designs
the term	4 weeks	1 year
the unit of money	1 thousand yen	1000 thousand yen

Table 7. Distribution of β shows that we get positive time preference rate for 3/4 subjects.

	0-0.31	0.31-0.4	0.41-0.45	0.46-0.5	0.51-
present experiment	5	5	32	32	24 (%)
experiment at WO	2	4	22	67	4 (%)

Table 8. Distribution of ρ of this experiment is larger. It seems come from the length of the term the subjects confront with their experiments.

	0.3-0.6	0.61-0.7	0.71-0.8	0.81-0.9	0.91-1
present experiment	0	3	14	49	35 (%)
experiment at WO	14	10	24	20	31 (%)

Table 9. Distribution of α shows more risk lovers were identified at the experiment of Wada and Oda [6]. In this experiment, subjects didn't need avoid risk because of the stakes of lottery are were small.

unit(per cent)	expected value of lottery Y	0-0.25	0.251-0.5	0.51-1	1.1-2	2-
present experiment	1000 yen	5	16	28	33	19 (%)
experiment at WO	1000 thousands yen	22	25	36	11	6 (%)

Table 10. Proportion of dynamically consistent choices shows the fact that a reasonable number of subjects are dynamically consistent in both experiment.

Confirmable	null hypothesis $\rho=\alpha$	t-test :(number)
present experiment	14/35=0.400	0.001068 rejected (10)
experiment at WO	20/43=0.465	0.008259 rejected (19)

References

1. Epstein, L. and S. Zin (1989) Substitution, Risk Aversion and Temporal Behavior of Consumption and Asset Returns: A Theoretical Framework, *Econometrica* 57:937-969
2. Epstein, L. and S. Zin (1991) Substitution, Risk Aversion, and the Temporal Behavior of Consumption and Asset Returns: An Empirical Analysis, *Journal of Political Economy* 99(2):263-86
3. Hall, R (1978) Stochastic Implications of the Life Cycle-Permanent Income Hypothesis: Theory and Evidence, *Journal of Political Economy* 86:971-987
4. Kreps, D. and E (1978) Porteus, Temporal resolution of uncertainty and dynamic choice theory. *Econometrica* 46:185-200
5. Mehra, R. and Prescott, E (1985) The equity premium: a puzzle, *Journal of Monetary Economics* 15:145-161
6. Wada, R. and Oda, S (2004) Separation of Intertemporal Substitusion and Preference Rate from Risk Aversion:Experimental Analysis, *in The proceedings of the Experiments in Economic Sciences: New Approaches to Solving Real World Problem. http://www.kyoto-su.ac.jp/project/orc/execo/EES2004/proceedings.html*

Signal Qualities, Order of Decisions, and Informational Cascades: Experimental Evidences

Sunichiro Sasaki

Osaka University

1 Introduction

Informational cascades are said to occur if players ignore private signals by following an established pattern of actions that their predecessors have chosen as a result of Bayesian updating in a sequential decision-making problem. Anderson and Holt [1] confirmed that informational cascades certainly occur in the laboratory as Bikhchandani et al. [2] suggests. However, other experimental studies including Çelen and Kariv [3], Huck and Oechssler [4], Kraemer et al. [5], and Nöth and Weber [6] generally argue that subjects put more weight on their private sig-nals than the Bayesian model assumes.

Following these results, this experimental study investigates the effects of signal precision on the formation of informational cascades by introducing heterogeneous signal qualities associated with the fixed order of deci-sions on the two different decision-making systems, anti-seniority and seniority.

2 Analytical Framework

There are two states of the world $\omega \in \{A, B\}$. Each state is realized equally likely, $Pr(A) = Pr(B) = 1/2$. In the experiment, each of six subjects $i \in \{1, 2, ..., 6\}$ does not observe the realized state, but receives a private signal σ^i_ω for the underlying state ω. The quality of private signal $Pr(A \mid \sigma^i_A)$ or $Pr(B \mid \sigma^i_B)$ is drawn from the six levels of precision $\{.55, .6, .65, .7, .75, .8\}$ and is exogenously determined by the position to which a subject is assigned. In addition to private signals, subjects later assigned to position 2 can observe their predecessors' predictions. After observing the private signal and the predecessors' predictions, each subject makes a prediction π^i_ω of which states would be realized one-by-one in sequence.

In the experiment, two treatments *anti-seniority* [1] and seniority are conducted. The difference in each treatment lies in the combination of signal

[1] The term anti-seniority is used in Ottaviani and Sørensen [7].

qualities and order in which predictions are made. In anti-seniority, six subjects make predictions in ascending order of the signal qualities. That is, the subject who has the least precise signal makes the prediction in position 1 ($Pr(A \mid \sigma_A^1) = Pr(B \mid \sigma_B^1) = .55$), the subject who has the second least precise signal makes the prediction in position 2 ($Pr(A \mid \sigma_A^2) = Pr(B \mid \sigma_B^2) = .6$), and in like manner, the subject who has the most precise signal makes the prediction in position 6 ($Pr(A \mid \sigma_A^6) = Pr(B \mid \sigma_B^6) = .8$). On the other hand, in seniority, six subjects make predictions in descending order of the signal qualities. That is, $Pr(A \mid \sigma_A^1) = Pr(B \mid \sigma_B^1) = .8, Pr(A \mid \sigma_A^2) = Pr(B \mid \sigma_B^2) = .75, ..., Pr(A \mid \sigma_A^6) = Pr(B \mid \sigma_B^6) = .55$.

In the experiment, the combination of signal qualities and order of decisions are common knowledge among all subjects. Thus, if subjects act as Bayesians, these two different treatments would create different behavioral patterns in the aggregate as follows.

In anti-seniority, complete informational cascades [2] occur if subjects in the first three consecutive positions make the same predictions. For example, the posterior probability that state A would be realized given that the subjects in the first three consecutive positions have predicted A and the subject in position 4 observes σ^4 is $Pr(A \mid \pi_A^1, \pi_A^2, \pi_A^3, \sigma_B^4) = .593$. In this case, the subjects in later than position 4 should ignore their private signals by following the established pattern of predictions. In seniority, the subject in position 2 should make the same predictions as the first subject, even when the private signal of the second subject does not correspond to the prediction of the first subject since $Pr(A \mid \pi_A^1, \sigma_B^2) = .571$. This leads to the result that subjects in a round should always make unanimous predictions regardless of whether they enter informational cascades or they truthfully reveal their own private signals.

3 Experimental Procedure

At the beginning of each round, the experimenter announced which treatment was to be conducted in order to make the combination of signal quality and the order of decision common knowledge. The experimenter drew one of the two cards from a box. On the card a letter, either "A" or "B" was printed and the letter on the card drawn represented the state of the world for that round. After confirming the letter, the experimenter then hid it from the subjects until the round was completed. Each subject then drew one of the six cards for determining the combination of the signal quality and the order of the decision.

Private signals were implemented by having the subjects draw one of 20 white or red marbles from a box. The white marbles represented state A and the red marbles represented state B. Different signal qualities were created by

[2] For the definition of complete informational cascades, see section 4.

varying the proportion of white and red marbles in a box given the realized state, treatment, and the assigned signal quality for each subject. Tables 1 and 2 show the combinations of marbles.

Table 1. Combinations of marbles- anti-seniority

Anti-seniority							
		Position (Signal quality)					
Realized State	Marbles	1 (.55)	2 (.6)	3 (.65)	4 (.7)	5 (.75)	6 (.8)
A	White	11	12	13	14	15	16
	Red	9	8	7	6	5	4
B	White	9	8	7	6	5	4
	Red	11	12	13	14	15	16

Table 2. Combinations of marbles- seniority

Seniority							
		Position (Signal quality)					
Realized State	Marbles	1 (.8)	2 (.75)	3 (.7)	4 (.65)	5 (.6)	6 (.55)
A	White	16	15	14	13	12	11
	Red	4	5	6	7	8	9
B	White	4	5	6	7	8	9
	Red	16	15	14	13	12	11

The experimenter approached each subject in turn and presented the box containing the exact proportion of white and red marbles for each subject from the set of six [3]. The subject drew a marble from the box and wrote down the state indicated by the color on the subject's record sheet. Then, the subject made a prediction by writing down one of the two states he or she thought which was more likely to be on the subject's record sheet. The experimenter wrote down the subject's signal and prediction on the experimenter's record sheet. The experimenter then approached the subject in the next position and showed the sequence of his or her predecessors' predictions. After all six subjects submitted predictions, the card the experimenter had drawn at the beginning of the round was revealed. This process was repeated 16 times in one session with combinations of each treatment.

[3] A set of six identical boxes, upon which codes were marked, was stored in a larger box separately from the states and kept in another room. After confirming the state, the experimenter brought the appropriate set from that room so that subjects could not identify which set was actually used. By checking the codes on the box, the experimenter could choose the appropriate box for each subject.

Sixty-six undergraduates at Keio University participated in the experiment. After the session, subjects were privately paid their payoffs in cash. For each correct prediction, 200 Japanese yen (equivalent to $1.84) were paid. The average payment was 2309 yen.

4 Results

In order to compare the aggregated behavior of each treatment, the analysis is based on the following three criteria: complete positive (negative) cascades, partial cascades, and full revelations.

A complete positive (negative) cascade, hereafter **CPC** (**CNC**), denotes a pattern of behavior such that at least one subject ignores his or her private signal by following the established pattern of predictions, and all of the six subjects in a round make unanimous correct (incorrect) predictions. Among **CPC** (**CNC**), a pattern of behavior such that all predictions are consistent with Bayesian posterior is called a Bayesian consistent complete positive (negative) cascade, hereafter **BCCPC** (**BCCNC**). A partial cascade, hereafter **PC**, denotes a pattern of behavior such that at least one subject ignores his or her private signal by following the established predictions, but at least one subject collapses it. A full revelation, hereafter **FR**, denotes a pattern of behaviors such that all of the subjects make predictions consistent with their private signals. Note that **CPC** (**CNC**), **PC**, **FR** do not overlap each other.

Result 1: Subjects made more correct predictions in seniority than in anti-seniority.

The observed proportions of correct predictions are higher in seniority than in anti-seniority (81.94% vs. 69.49%). The Mann-Whitney U test shows that this difference is statistically significant (z=-4.335).

Result 2: ***CPC, CNC, BCCPC, and BCCNC*** *occurred more frequently in seniority than in anti-seniority, whereas* ***PC*** *occurred more frequently in anti-seniority than in seniority.*

The observed proportions of **CPC** are higher in seniority than in anti-seniority (60.00% vs. 14.29%), of which difference is significant (z=-6.204). The observed proportions of **CNC** are higher in seniority than in anti-seniority (10.00% vs. 1.79%), of which difference is significant (z=-2.431). The observed proportions of **BCCPC** are higher in seniority than in anti-seniority (60.00% vs. 6.25%), of which difference is significant (z=-7.736). The observed proportions of **BCCNC** are higher in seniority than in anti-seniority (10.00% vs. .89%), of which difference is significant (z=-2.873). The observed proportions of **PC** are higher in anti-seniority than in seniority (25.89% vs. 10.00%), of which difference is significant (z=2.460).

Result 3: ***FR*** *occurred more frequently in anti-seniority than in seniority.*

The observed proportions of **FR** are higher in anti-seniority than in seniority (58.04% vs. 20.00%), of which difference is significant (z=4.767).

Result 4: For both treatments, ***BCCPC*** *and* ***BCCNC*** *occurred less frequently than the Bayesian theory predicts.*

The probability that **BCCPC** (**BCCNC**) would occur is given by the probability that unanimous predictions by **FR** occur, subtracted from the probability that **CPC** (**CNC**) occur since unanimous predictions by **FR** are not counted in **BCCPC** (**BCCNC**). In anti-seniority, **BCCPC** for state A[4] occur with $[Pr(A \mid \sigma_A^1) \times Pr(A \mid \sigma_A^2) \times Pr(A \mid \sigma_A^3)] - [Pr(A \mid \sigma_A^1) \times Pr(A \mid \sigma_A^2) \times Pr(A \mid \sigma_A^3) \times Pr(A \mid \sigma_A^4) \times Pr(A \mid \sigma_A^5) \times Pr(A \mid \sigma_A^6)] = .125$. By the same calculation, **BCCNC** for state A occur with .062. In seniority, **BCCPC** for state A occur with $Pr(A \mid \sigma_A^1) - [Pr(A \mid \sigma_A^1) \times Pr(A \mid \sigma_A^2) \times Pr(A \mid \sigma_A^3) \times Pr(A \mid \sigma_A^4) \times Pr(A \mid \sigma_A^5) \times Pr(A \mid \sigma_A^6)] = .710$. By the same calculation, **BCCNC** for state A occur with .199.

However, observed proportions of **BCCPC** and **BCCNC** are lower than the probabilities calculated above (for the observed proportions, see Result2). The one-tailed population proportion tests show that for both treatments, the observed proportions of **BCCPC** and **BCCNC** are significantly lower than the theoretical predictions (z=-2.000 in anti-seniority, z=-1.878 in seniority for **BCCPC** and z=-2.329 in anti-seniority, z=-1.921 in senior-ity for **BCCNC**).

5 Conclusion

In a series of experiments we observed the different consequence of two treatments where signal precision and order of decision are different. Our results suggest that **CPC** and **CNC** occur more frequently in seniority than in anti-seniority, that seniority is more efficient than anti-seniority, but increases the risk of creating **CNC**, that private signals can be extracted more effectively in anti-seniority than in seniority. By calculating the probability **BCCPC** and **BCCNC** would occur, we observe clearly that human subjects do not always behave as the Bayesian theory predicts.

References

1. Anderson, L. R. and C. A. Holt (1997) "Information Cascades in the Laboratory" American Economic Review 87, 847-62.
2. Bikhchandani, S., D. Hirshleifer, and I. Welch (1992) "A Theory of Fads, Fashion, Custom, and Cultural Change as Informational Cascade" Journal of Political Economy 100, 992-1026.
3. Çelen, B. and S. Kariv (2004) "Distinguishing Informational Cascades from Herd Behavior in the Laboratory" American Economic Review 94, 484-98.

[4] We consider only the case for state A because of symmetric posteriors.

4. Huck, S. and J. Oechssler (2000) "Informational Cascades in the Laboratory: Do They Occur for the Right Reasons?" Journal of Economic Psychology 21, 661-71.
5. Kraemer, C., M. N'oth, and M. Weber (2000) "Information Aggregation with Costly Information and Random Ordering: Experimental Evidence" Sonder Forschungs Bereich 504 Working Paper Series 00-35, Universi'at Mannheim.
6. Nöth, M. and M. Weber (2003) "Information Aggregation with Random Ordering: Cascades and Overconfidence" Economic Journal 113, 166-189.
7. Ottaviani, M. and P. N. Sørensen (2001) "Information Aggregation in Debate: Who Should Speak First?" Journal of Public Economics 81, 393-421.

WTP and WTA for Expressway Services

Metin Senbil[1] and Ryuichi Kitamura[2]

[1] Hiroshima University
[2] Kyoto University

1 Introduction

Changes in consumer surplus have been approached in two different ways [1,2]: one is the compensating variation (CV), which is the amount of income that an individual is ready to pay to keep his utility as it was before a change; the other is the equivalent variation (EV), which is the amount of money the individual is ready to accept for the change. For welfare gains, CV and EV are known as willingness to pay (to attain the gain, WTP) and willingness to accept (to accept the absence of the gain, WTA) respectively; but for a welfare loss, CV and EV refer to WTA (to compensate the loss) and WTP (to prevent the loss).

In the last thirty years or so, it has been shown that WTP and WTA are not equal to each other [3]. Substitution effect [4,5], income effect [6], loss and gain disparity [7,8], endowment effect [9], property rights [10], and transaction costs [11] have been found to account for the WTA/WTP disparity.

Transportation studies have been concerned with the value of travel time (VOTT), which can be viewed as WTP for additional time savings. There has been a vast literature on VOTT accumulated over time, which will not be reviewed here. Interested reader is referred to [12] and [13]. To the knowledge of the authors, most of the revealed-preference or the stated-preference studies on VOTT derive VOTT values from estimated indirect utility functions through the trade off between cost and time embedded in the utility function. Studies by [12], [14], [15], and [16] are examples where this approach has been used. A different approach [17] employs conjoint analysis, and estimates VOTT directly by including the toll into the disutility function.

Although travel time is a pivotal element in all of the VOTT studies, other constraints such as time pressure on commuters exist, and yet have not been well represented. It might be the case that WTP is substantially affected by time pressure, which arises from the presence of a coupling time (the time one must report at the destination), and the uncertainty caused by travel time variability. In this regard, WTP might not be a static measure as depicted in previous studies, but rather it may be a dynamic quantity whose value can change according to the coupling time, time of day, and the nature of the activities at the destination.

It can be argued that incorporating the notion of coupling time into the quantification of WTP facilitates the quantification of WTA as a quantity distinct from WTP. Equally important, a commuter might be more sensitive to a definite loss in time caused by, e.g., arriving too early, than to travel time variation. To test these hypotheses, we have conducted a survey which contained stated-preference questions of the choice between a toll road and surface streets under a coupling time and with travel time variations. The following is a brief description of the survey and its preliminary results.

2 Survey

It is postulated in this study that trips on surface streets are the default choice for commuters, but bear larger degrees of risk of tardiness; because of this reason, a commuter would choose to switch from surface streets to the expressway at the payment of a toll. In the survey, the respondent is asked to report how much he is willing to pay for the expressway in open-ended format. In order to derive WTP values, we adopt the reference time point approach [18] which has been derived from prospect theoretic notions [8]. Two reference time points are postulated: the earliest possible arrival time (EAT), i.e., the earliest time a commuter can arrive, and the latest acceptable arrival time (LAT), which is assumed to be the coupling time presented in the survey question. Arrivals later than LAT are indexed as losses, and arrivals before LAT as gains. In the survey, seven questions were asked to the respondents, each having a different coupling time and a different travel time variation on the surface street alternative. In all of the questions, respondents are presented with an identical travel time variation for the expressway. Table 1 presents each of the questions, expressed in terms of the length of the band of possible arrival times by the surface streets in the gain and loss regions.

A total of 222 respondents out of 236 responded to the WTP questions. Mean WTPs for the respective questions show a certain degree of variability, but other summary statistics such as modes and medians show extreme regularity, saddling at ¥ 500 for all questions. It might be the case that the responses bear some rounding errors toward ¥ 500. When faced with a small gain interval, respondents are risk aversive and increase their WTP. It can be said that WTP for expressways increases when the risk of tardiness on surface streets increases.

An inspection of Table 1 would show, Questions 1 and 2 present large probabilities of early arrival by surface streets. Questions 3 and 4 represent cases where the arrival time band by surface streets is centered at the reference point, with Question 4 having a band which is five minutes longer on each

Table 1. Arrival time band width in the gain and loss regions by surface streets

Question#	Gain Region	Loss Region
1	15	5
2	25	5
3	10	10
4	15	15
5	5	15
6	5	25
7	0	20

The arrival band width in the gain region is determined as (coupling time) - (earliest possible arrival time), and the one in the loss region as (latest possible arrival time) - (coupling time), and expressed in minutes.

side. Questions 5 and 6 offer arrival time bands which are symmetrical, with respect to the reference point, of those of Questions 1 and 2, respectively.

Note that an in-time arrival is impossible with the surface streets in Question 7. We might say that the value reported for the expressway in Question 7 represents an upper bound of WTA. In all of the questions except Question 7, respondents have presumably reported values by assessing the losses in the face of gains. On the other hand, there are no possibilities of gains by the surface streets in Question 7; the expressway is the only possible way for an in-time arrival, which respondents must use to secure an in-time arrival. In this respect the value that they report becomes EV or WTA for arriving in time.

One may notice that questions can be paired together according to the earliest or latest arrival time by the surface streets. For example, question pairs 4-5 and 1-2 have different gains but the same losses. Question pairs 1-4 and 5-6 display different losses and the same gains. This allows us to analyze variations in reported WTPs resulting from a difference only in the gain or loss region, and consequently to check their consistency with the prospect theoretic value function [8]. As depicted in Figure 1, the value function is concave for gains and convex for losses, and steeper for losses than for gains. Figure 1 also shows the gain and loss bands of the respective questions.

The comparison of the paired differences between L1 and L2 or equally G1 and G2 as defined in Figure 1 would indicate the behavior of the value function for losses and gains of different magnitudes. According to the postulates of the value function, the following alternative hypotheses can be derived from Figure 1:

1. L1 > L2: due to the convexity of the value function in the loss region
2. G1 > G2: due to the concavity of the value function in the gain region
3. L1 > G1: due to the steeper slope of the value function in the loss region
4. L2 > G2: due to the steeper slope of the value function in the loss region

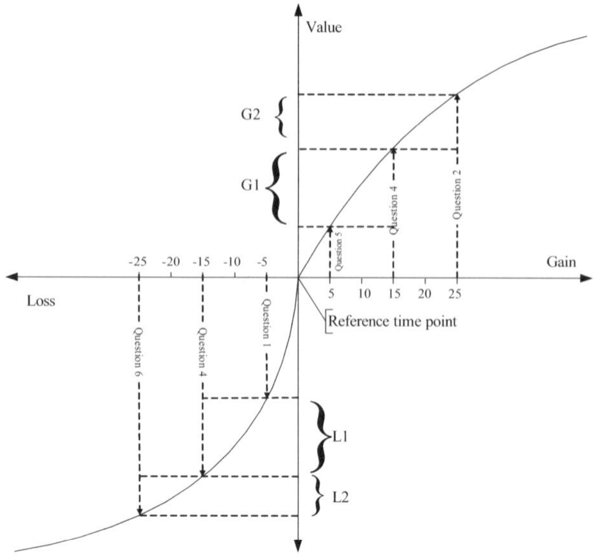

Fig. 1. Theoretical value function

To test the validity of these alternative hypotheses, we have conducted paired t-tests. We found that L1 is not significantly larger than L2 ($p = 0.19$). In the gain region, it is likewise found that G1 and G2 are not different from each other ($p = 0.97$). On the other hand, when we compare gains and losses, we find that L1 is significantly larger than G1 ($p = 0.00$). The same is true for the comparison between L2 and G2: L2 is significantly larger than G2 ($p = 0.00$).

The reason why G1 and G2 as well as L1 and L2 are not very different from each other may be attributed to the relative location of the expressway arrivals to the reference time point. Because the simple comparisons made above assume that there is no variation in the values the expressway alternative would produce.

Figure 2 summarizes the information given in Table 1 in a diagrammatical way. By inspecting Figure 2 closely, one can notice that expressway arrival times are different across questions with respect to the reference time point. For only two question pairs, Questions 6-7 and Questions 3-4, the arrival times by the expressway fall in the same locations relative to the reference time point.

Theoretically speaking, WTP for Question 4 should be larger than that for Question 3 as the value function is steeper in the loss region than the same interval in the gain region. This is supported by a paired t-test that indicates that the mean value reported for Question 4 is significantly larger than that for Question 3 ($p = 0.01$). Likewise, the mean WTP for Question 7 is significantly larger than that for Question 6 ($p = 0.01$), which is not

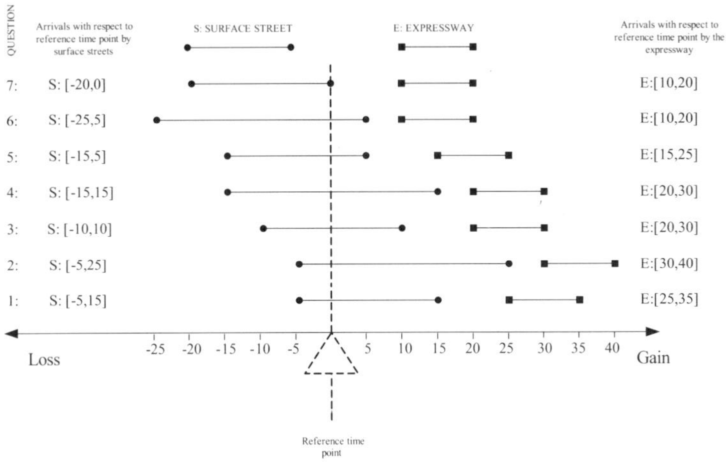

Fig. 2. Gains and losses in the seven survey questions

possible in prospect theory as a five-minutes loss should be valued more than a five-minutes gain.

References

1. Hicks JR (1943) The four consumer surpluses. Review of Economic Studies 8: 108-116
2. Hicks JR (1956) A Revision of Demand Theory. Clarendon Press, Oxford, UK.
3. Hammack J, and Brown GM (1974) Waterfowl and Wetlands: Towards Bioeconomic Analyses. Johns Hopkins University Press, Baltimore, MD.
4. Hanemann MW (1991) Willingness to pay and willingness to accept: how much can they differ? American Economic Review 81: 635-647
5. Shrongen JF Shin, S.Y. and Hayes, D.J. (1994) Resolving differences in willingness to pay and willingness to accept. American Economic Review 84: 255-269
6. Willig RD (1976) Consumer's surplus without apology. American Economic Review 66: 589-597
7. Brookshire DS, and Coursey DL (1987) Measuring the value of a public good: an empirical comparison of elicitation procedures. American Economic Review, 77: 554-566
8. Kahneman D, and Tversky A (1979) Prospect theory: an analysis of decision under risk. Econometrica 47: 263-271
9. Thaler R (1980) Toward a positive theory of consumer choice. J Economic Behavior and Organization 1: 39-60
10. Coase R 1960. The problem of social cost. J Law and Economics 3: 1-44
11. Brown TC, and Gregory R (1999) Why the WTA-WTP disparity matters. Ecological Economics 28: 323-335
12. Small KA (1992) Urban transportation economics. Fundamentals of Pure and Applied Economics Series, 51. Harwood Academic Publishers, New York.

13. Wardman M (2001) A review of British evidence on time and service quality valuations. Transportation Research E 37: 107-128
14. Lam TC, and Small KA (2001) The value of time and reliability: measurement from a value pricing experiment. Transportation Research E 37: 231-251
15. Mackie PJ, Jara-Diaz S, and Fowkes AS (2001) The value of travel time savings in evaluation. Transportation Research E 37: 91-106
16. Brownstone D, Ghosh A, Golob TF, Kazimi C, and Van Amesfort D (2003) Drivers' willingness-to-pay to reduce travel time: evidence from the San Diego I-15 congestion pricing project. Transportation Research A 37: 373-387
17. Calfee J, and Winston C (1998) The value of automobile travel time: implications for congestion policy. J Public Economics 69: 83-102
18. Senbil M, and Kitamura R (2004) Reference points in commuter departure time choice: a prospect theoretic test of alternative decision frames. Journal of Intelligent Transportation Systems 8: 19-31

Judgement in Small Decision-Making Problems

Takemi Fujikawa[1] and Sobei H. Oda[2]

[1] University of Western Sydney
[2] Kyoto Sangyo University

1 Introduction

This paper experimentally investigates "small decision-making" (SDM) problems on the ground that many common activities include those problems. SDM problems are defined by three main properties [2]. First, they include repeated tasks; the decision makers (DMs) face the same choice problem many times in similar situations. Second, each single choice is not very essential; the alternatives tend to have similar expected value (EV) that may be fairly small. Finally, the DMs do not have objective prior information as to payoff distribution. In choosing among the possible alternatives, the DMs will have to rely on the immediate and unbiased feedback obtained in similar situations in the past.

This paper proposes that the exploration tendency in SDM problems cannot be explained by the DMs' risk attitude. Rather, each DM's behaviour in SDM problems can be justified by the *search-assessment model*, which is a key contribution to this paper. The model insists that in SDM problems the tendency to select best reply to the past, and misestimation of payoff distribution can lead to robust deviations from EV maximisation.

We start an analysis by reviewing Barron and Erev [2]'s experimental research on SDM problems. They reported the results revealing differences in SDM problems as opposed to conventional "big decision-making" (BDM) problems. The experiment included the treatment in which the choice situation was repeated many times, the alternatives had similar and small EV, and the distribution of lotteries was unknown to the subjects. The search-assessment model gives a key for improving our understanding of the behavioural tendency reported by [2].

The rest of this paper is structured as follows. Section 2 reviews Kahneman and Tversky [7]'s study. Section 3 reviews Barron and Erev [2]'s experiment on SDM problems. We present the search-assessment model in Section 4. Finally we conclude.

2 A Previous Experiment on BDM Problems

Allais [1] maintained that the more risky of two prospects became relatively more attractive when the probability of winning in both prospects was multiplied by a common ratio. This "common ratio" effect is of great importance since it represents a robust violation of the tenets of expected utility theory (EUT) [11]. To ascertain the validity of the common ratio effect, Kahneman and Tversky [7] performed the following pair of choice problems (the outcomes represented hypothetical payoffs in thousand Israeli Lira):

Problem A. Choose between:
H: 4 with probability 0.8 ; 0 otherwise N=95
L: 3 with certainty
Problem B. Choose between:
H: 4 with probability 0.2 ; 0 otherwise N=95
L: 3 with probability 0.25 ; 0 otherwise.

Kahneman and Tversky [7] showed that while 80% of their subjects preferred L in Problem A, only 35% preferred L in Problem B. However, their results revealed a violation of the independence axiom of EUT that asserts that the DMs should have the same preferences in the two problems. The comparison of Problem A and B described a series of choice problems in which the DMs' preferences systematically violated such axiom. The results showed that the DMs overweighted outcomes that were considered certain, relative to the outcomes which were merely probable–a phenomenon which was labelled the *certainty effect*.

3 A Previous Experiment on SDM Problems

Barron and Erev [2] replicated Problem A and B, and reported the results revealing differences in risk attitude in SDM problems as opposed to conventional BDM problems conducted by a number of researchers (e.g. [1], [3], [7], [8], [9], [10]). As well as Problem A and B, Problem C was conducted to examine *underweighting of rare events*:

Problem C. Choose between:
H: 32 with probability 0.1 ; 0 otherwise
L: 3 with certainty.

In [2], the DMs' task was to choose either H or L 400 times in each of Problem A, B and C. For example, one selection of H in Problem C made the DMs earn 32 points with probability 0.1 and zero point with 0.9. The information available to the DMs was limited to feedback concerning the outcomes of their previous decisions. Note that in [7], the DMs were correctly disclosed payoff distribution and they performed only one round with a hypothetical payoff. On the other hand, the DMs in [2] were not disclosed payoff distribution, asked to choose 400 times and paid real money according to their performance at a conversion rate of one point to 0.01 Israeli Shekel (0.25 US cent).

Barron and Erev [2] claimed the following with experimental results. First, they claimed the observation of the *reversed* certainty effect. While the mean proportions of H choices (*choiceH*) over the subjects were 0.63 for Problem A, it decreased significantly to 0.51 for Problem B. Second, the observation of underweighting of rare events (i.e. 32 with p=0.1) was claimed with the result that *choiceH* in Problem C was 0.28. However, it is not safe to make direct comparison between the results in [2] and [7] due to the following reason.

It was not examined whether or not the subjects in [2] could correctly estimate each alternative only with hundreds of trials. As they did not have prior information as to payoff distribution, the subjects would have to refer to feedback of their past outcome in every round to estimate payoff distribution. In the process of trying alternatives repeatedly, the subjects would gradually form their subjective payoff distribution, which was or was not the same as objective payoff distribution. Having finished searching for payoff distribution of both alternatives (H and L) with only 400 trials, some subjects regarded H (L) as the alternative with higher (lower) EV, that is, they had estimated the alternative correctly. Others, however, would regard H (L) as the alternative with lower (higher) EV, that is, they had misestimated the alternative. Therefore, it is quite ambiguous that the subjects in [2] chose H (L) supposing it had higher (lower) EV.

4 The Search-Assessment Model

We have pointed out that the DMs were likely to misestimate the payoff structure of binary choice problems in [2]. To demonstrate that the probability of such misestimation is rather high, we will propose the *search-assessment model*.

Suppose that the DM is asked to choose either H or L in Problem D at each round t $(t = 1, \ldots, 400)$:
Problem D. Choose between:
$H : x \quad (p) \qquad ; \quad 0 \quad (1 - p)$
$L : 1 \quad (1)$,
where $p \in (0,1)$ and x such that $px > 1$. The DM gets x points with probability p and zero point with $(1 - p)$ by choosing H; she/he gets one point for sure by choosing L. Note that EV of H $(EV(H))$ is greater than EV of L $(EV(L))$. To attempt an analysis of Problem A and C, we examine Problem D, which applies to Problem A by setting p=0.8 and x=4/3, and applies to Problem C by setting p=0.1 and x=32/3. We here define the posterior average points of H (*posteriorH*) as the points the DM has earned from H after it was chosen $m(0 \leq m \leq 400)$ times. For example, if the DM has acquired 12 points from H after she/he chose it five times, then *posteriorH* is 2.4 (=12/5).

We denote $P(H_m)$ as the probability of the event that the DM's *posteriorH* becomes greater than or equal to one that is $EV(L)$ after m selections of H:

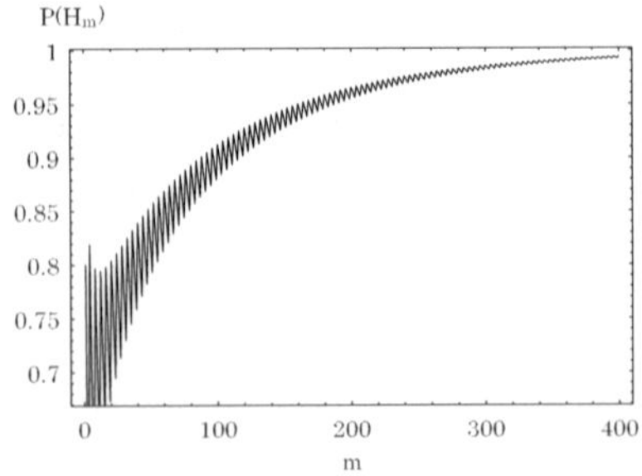

Fig. 1. The probability of judging that $EV(H) \geq EV(L)$ after m selections of H in Problem A

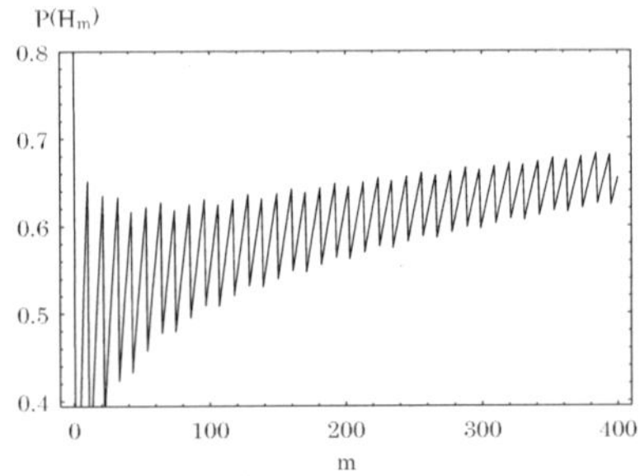

Fig. 2. The probability of judging that $EV(H) \geq EV(L)$ after m selections of H in Problem C

$$P(H_m) = \sum_{\text{all } m:\, \frac{kx}{m} \geq 1} {}_mC_k\, p^k\, (1-p)^{m-k} = \sum_{k=\lfloor \frac{m}{x} \rfloor + 1}^{m} {}_mC_k\, p^k\, (1-p)^{m-k}. \quad (1)$$

Put differently, one regards $P(H_m)$ as the probability of the event that the DM can judge that $EV(H) \geq EV(L)$ after she/he has chosen H m times.

Calibration of $P(H_m)$ allows us to analyse an optimal number of H choices needed for judging that H has higher EV than L. Fig. 1 – 2 depict $P(H_m)$ for Problem A and C respectively. Fig. 1 shows that 0.97 is the probability of the event that *posteriorH* exceeds three after 200 selections of H in Problem A, that is, the DM may have judged that $EV(H) \geq EV(L)$, *ex post*. Similarly, Fig. 2 shows that 0.63 is the probability of the event that *posteriorH* exceeds three after 200 selections of H in Problem C. Interestingly, $P(H_m)$ does not exceed 0.98 until the DM chooses H 10,000 times in Problem C.

Next, let us examine the following Problem E to analyse Problem B, where both of the two alternatives, H and L, include uncertain outcomes:

Problem E. Choose between:

$H : x \quad (\theta p) \quad ; \quad 0 \quad (1-\theta p)$

$L : 1 \quad (\theta) \quad ; \quad 0 \quad (1-\theta)$,

where p, $\theta \in (0,1)$ and x such that $\theta p x > \theta$. The DM in Problem E is to be instructed to choose either H or L at each round t $(t = 1, \ldots, 400)$. One selection of H makes the DM earn x points with probability θp and zero point with (1-θp); one selection of L makes the DM earn one point with probability θ and zero point with (1-θ). Problem E applies to Problem B by setting p=0.8, θ=0.25 and x=4/3.

We now define $P(L_n)$ as the probability of the event that *posteriorH* becomes greater than or equal to *posteriorL* after the DM has chosen H and L m and n times in Problem E respectively:

$$P(L_n) = \sum_{k=0}^{m} \left[{}_mC_k \, (\theta p)^k (1-\theta p)^{m-k} \times \sum_{j=0}^{\left\lceil \frac{nkx}{m} \right\rceil} {}_nC_j \, (\theta)^j \, (1-\theta)^{n-j} \right], \quad (2)$$

where $n \in N$ such that $0 \leq m + n \leq 400$. Note that if the DM chooses H m times and gets x points k times, then her/his *posteriorH* is kx/m; if the DM chooses L n times and gets 1 point l times, then her/his *posteriorL* is l/n. Calibration of $P(L_n)$ tells us that if the DM chooses H and L each 200 times in Problem B, 0.64 is the probability for an event that she/he can judge that $EV(H) \geq EV(L)$ as shown in Fig. 3.

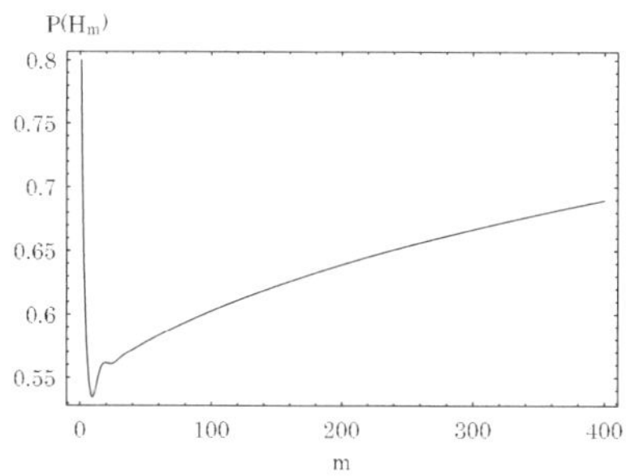

Fig. 3. The probability of judging that $EV(H) \geq EV(L)$ after m selections of H in Problem B

5 Conclusion

We have investigated SDM problems in the context of binary choice problems. For this investigation, we first reviewed Barron and Erev [2]'s experiment on SDM problems. The subjects were asked to choose one of two options 400 times: (1) the risky option with higher EV and (2) the safe option with lower EV. Recall that the subjects were not disclosed payoff structure during trials. The results revealed that 28% of the subjects preferred H to L in Problem C. It claimed that the subjects revealed deviations from EV maximisation, and justified that such deviations would result from underweighting of rare events.

A key contribution to this paper includes the search-assessment model, which offers better explanation of Barron and Erev [2]'s results, rather than the assertion of underweighting of rare events they insisted. As the subjects in [2] were not disclosed payoff structure, some subjects would have misestimated payoff structure. The search-assessment model asserts that the subjects are very likely to misestimate the payoff structure with only 400 trials. Specifically, the model asserts that 0.63 is the probability for the event that the subjects can understand correctly the payoff structure of the binary choice problem (i.e. Problem C), even if they have performed the alternative all the time. The event does not have probability of 0.98 until the subjects perform 10,000 trials. Therefore, it is quite ambiguous that Barron and Erev [2]s' subjects chose H (L) supposing it has higher (lower) EV.

This paper has proposed that the exploration tendency in SDM problems seems to be explained by the search-assessment model, rather than each DM's risk attitude. The model has insisted that in SDM problems the tendency to select best reply to the past, and misestimation of payoff distribution can lead to robust deviations from EV maximisation.

References

1. Allais M (1953) Le Comportement de l'Homme Rationnel devant le Risque: Critique des Postulats et Axiomes de l'Ecole Americaine. Econometrica 21:503–546
2. Barron G, Erev I (2003) Small Feedback-Based Decisions and Their Limited Correspondence to Description-Based Decisions. Journal of Behavioral Decision Making 16:215–233
3. Conlisk J (1989) Three Variants on the Allais Example. American Economic Review 79:392–407
4. Erev I (2004) On the Weighting of Rare Events and the Economics of Small Decisions. Proceedings of the International Conference, Experiments in Economic Sciences: New Approaches to Solving Real-world Problems
5. Erev I, Barron G (2005) On Adaptation, Maximization, and Reinforcement Learning Among Cognitive Strategies. Psychological Review 112:912–931
6. Fujikawa, T (2005) An Experimental Study of Petty Corrupt Behaviour In Small Decision Making Problems. American Journal of Applied Sciences Special issue:14–18
7. Kahneman D, Tversky A (1979) Prospect theory: An Analysis of Decision under Risk. Econometrica 47:263–291
8. Levi, I (1986) The Paradoxes of Allais and Ellsberg. Economics and Philosophy 2:23–53
9. Machina, M (1982) "Expected Utility" Analysis without the Independence Axiom. Econometrica 50:277–324
10. Schmidt, U (2000) The Certainty Effect and Boundary Effects with Transformed Probabilities. Economics Letters 67:29–33
11. Von Neumann, J, Morgenstern, O (1944) Theory of Games and Economic Behavior. New Jersey: Princeton University Press

Part IV

Game-Theoretic Individual Decision Making

The Effect of Inter-group Competition in the Prisoner's Dilemma Game*

Yoshio Iida

Kyoto Sangyo University

1 Introduction

This study focuses on the effect of inter-group competition in the prisoner's dilemma game. Recent experimental studies have investigated inter-group competitions in which matched teams compete against each other to produce larger contributions (cooperation within one's group). Bornstein et al. and Goren (Bornstein [3], Bornstein and Ben-Yossef [2], Bornstein, Erev and Goren [3], Goren [6], Goren and Bornstein [6]) investigate the inter-group competition of a special prisoner's dilemma. Features of their game fit the examples of lobbying, wars, and similar forms of competition: 1) no contribution is a dominant strategy for all players in the game, 2) no contribution is the collectively (i.e., Pareto) efficient outcome of the game. The studies also show that competition has a positive effect on the cooperative decision, but that the effect does not last when the game is repeated.

In this study I focus on the effect of competition in a very normal PD game in which defection is the dominant strategy for each player and the cooperation of both players brings about a Pareto-efficient outcome. My purpose is to investigate how competition works as a mechanism to help an economy achieve a Pareto-efficient outcome. By conducting an inter-group competition game with the stakes set at several levels, this study seeks to identify the factors required for reaching a collectively efficient outcome.

Nalbantian and Schotter [7] studied an inter-group competition in an iterated normal public goods game. They set the stakes of their game sufficiently high to ensure that the dominate strategy would bring about the Pareto-efficient outcome. In their setting, subjects consistently made high-level contributions until the end of the game. However, people often invest great effort

* This research was supported by Japan Society for the Promotion of Science(JSPS), Grant-in-Aid for (B) No. 15730145 and the Open Research Center "Experimental Economics: A New Method of Teaching Economics and the Research on Its Impact on Society", Kyoto Sangyo University

to win a competition even when the gain or loss is somewhat paltry. The instinct to win and avoid loss may induce cooperative behaviors in an inter-group competition of a PD or PG game, even when the stakes are too low to compel the subjects to cooperate as a rational choice. To investigate whether this instinct is operative, I studied the inter-group competition in a prisoner's dilemma game played for high stakes, moderate stakes, and zero stakes. To distinguish the effect of stakes and that of information on the result of the competition (win or loss), I also conducted a normal PD game as a benchmark treatment.

The experimental results show the following: i) subjects cooperate if the stakes are high enough; ii) players maintain a high level of cooperation even if the stakes are moderate, hence betrayal is the dominant strategy for every player; iii) no significant difference is found between the high- and moderate-stake games even though their dominant strategies differ; iv) the players cooperate significantly more in high- and moderate-stake games than they do in the zero-stake game; and v) the players in the zero-stake game cooperate significantly more than they do in the benchmark treatment.

2 The Model

The experimental subjects (players) were assigned to two-person groups. Each player was given 1 token at the beginning of each period and asked to decide whether he would keep the token (a "betray strategy" in the prisoner's dilemma game) or invest it in a group project (a "cooperate strategy"). Subjects earned 10 points for each token they kept for themselves and 6 points for each token that they and their fellow group members contributed to the group project.

Two groups were matched and competed for stakes. Members of the group which collected the biggest investment won the stakes. The sources of the stakes were provided endogenously; that is, the group members who collectively invested less in their group project had to pay the stakes. The stakes were the same for all of the players. If the groups tied, no stakes were paid. A subject's payoff function was as follows:

$$U_{mi} = \begin{cases} 10(E_{mi} - g_{mi}) + 6\sum_{i=1}^{2} g_{mi} & if \sum_{i=1}^{2} g_{mi} = \sum_{i=1}^{2} g_{ni} \\ 10(E_{mi} - g_{mi}) + 6\sum_{i=1}^{2} g_{mi} + T & if \sum_{i=1}^{2} g_{mi} > \sum_{i=1}^{2} g_{ni} \\ 10(E_{mi} - g_{mi}) + 6\sum_{i=1}^{2} g_{mi} - T & if \sum_{i=1}^{2} g_{mi} < \sum_{i=1}^{2} g_{ni} \end{cases} \quad (1)$$

where E_{mi} is the initial endowment, g_{mi} is a voluntary contribution to the group project. The stake, T, is set at three levels, $T = 5$, $T = 3$, and $T = 0$, and the corresponding games (phases) are referred to as "GC5," "GC3," and "GC0". The dominant strategies are to contribute one's token to the group project in GC5 (cooperation) and to keep it in GC3 and GC0 (betrayal).

There were no stakes in GC0, but it differed from a normal PD game in that the players were informed of the amount of investment made by the opponent group. The players could use this information to compare their own group with their opponent group. In the Benchmark treatment, the subjects played a normal PD game. In the session, subjects had no interaction with the other groups and no information about the other groups during any session. Each game consisted of 20 rounds (game periods). The group matching and group members remained unchanged during the experiments (partner treatment). Students from Kyoto Sangyo University were hired to play the game series. 44 students participated in GC5, 48 in GC3, 56 in GC0, and 28 in the benchmark treatment. The experiments were conducted in a computerized laboratory with z-Tree software.

3 Results

3.1 Outline of Results

Figure 1 shows the results of the treatments. The average contribution rate over the 20 periods of GC5 was 83.4%. The subjects remained virtually rational throughout the entire phase except the first few periods. The average contribution rate over the 20 periods of GC3 was 74.1%. The contribution rate did not decrease from period to period, though contribution was not a dominant strategy. The contribution rate over the entire session of GC0 was 45.1%. The average contribution of each period decreased gradually as the end of the game approached.

Figure 1 also shows the results of the benchmark treatment. The average contribution over the 20 periods of the session was 24.8%. The decreasing trend was similar to that of GC0, but the contribution level was lower than that of GC0.

In visual inspection, the group average of GC5 and GC3 is clearly higher than the group averages of GC0 and the benchmark treatment. There also seems to be a significant difference between GC0 and the benchmark treatment.

Table 1 shows U values of the pair-wise robust rank order test between the treatments. The average contribution of GC5 is slightly higher than that of GC3, but the pair-wise robust rank order test accepts the null hypothesis at both the 5% and 10% significance levels. This means that there is no significant difference between the strategies even though the dominant strategies are different.

The null hypothesis is rejected at the 5% level of significance between GC5 and GC0 and between GC3 and GC0. This means that the contribution level of GC3 is significantly higher than that of GC0 even though the dominant strategies of these treatments are the same.

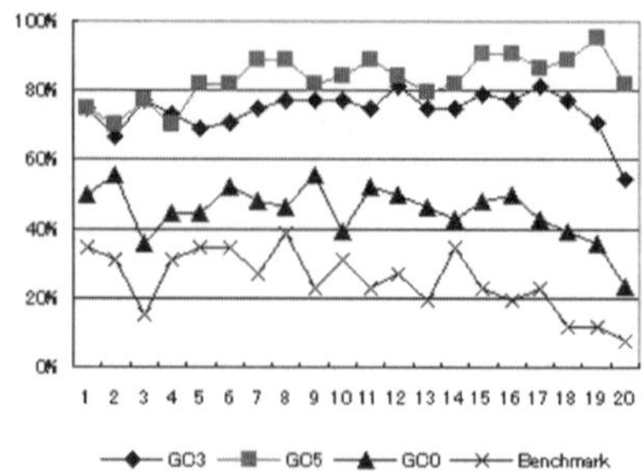

Fig. 1. Average Contribution

Table 1. U-values of Robust Rank Order Test

	GC3 74.2%	GC0 45.1%	Benchmark 24.8%
GC5 83.4%	0.73	3.61**	3.97**
GC3 74.2%	-	2.88**	3.88**
GC0 45.1%	-	-	2.90**

A significant difference is also found between GC0 and the Benchmark Session. The dominant strategies of these sessions are identical to each other, just as they are in the case of Sessions GC3 and GC0. The only difference between the game treatments is the availability of information on the opponent group.

3.2 Individual Decisions

Individual data shows that many of the subjects continued to contribute during the games in GC3 and GC5. The number of subjects who contributed more than 80% of periods was Twenty-seven (56.3%) in GC3, 31 (70.5%) in GC5, 11 (19.6%) in GC0 and 0 (0%) in the Benchmark Session.

Individual data also tells us that individual contribution rates within the groups were very similar. Figure 2 shows the differences between individual contribution rates in the groups. Most of the differences were less than 20% in any single game. The small differences in GC5 and GC3 are well explained by the high concentration of individual contributions rates within the range from 0.8 to 1. Yet similarly small differences appeared in GC0 and the Benchmark Session, games in which the individual contribution rates were widely distributed. Did the matched groups share similar contribution rates to the same degree as the members within the groups? Figure 3 shows the differences in the group contribution rates between the matched groups. The differences were small in most cases in GC5 and GC3 but not in GC0. As in the case observed within the groups, many of the group contribution rates of GC5 and GC3 were high. Thus, there could not be large differences in many of the cases. This result suggests that there is no considerable correlation between group average contributions of matched groups in the case of no stakes.

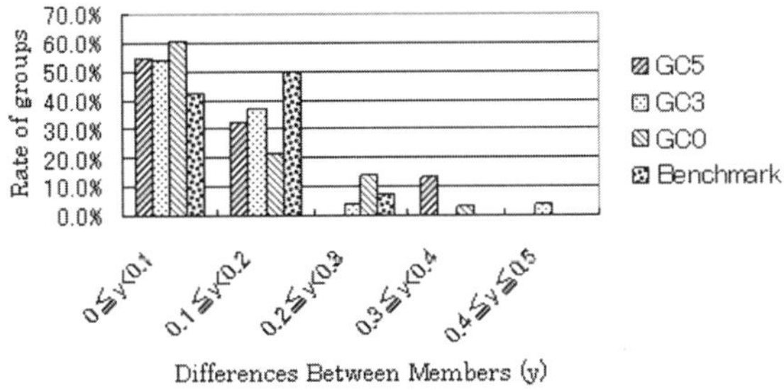

Fig. 2. Distribution of Differences between Individual Contribution Rates in Groups

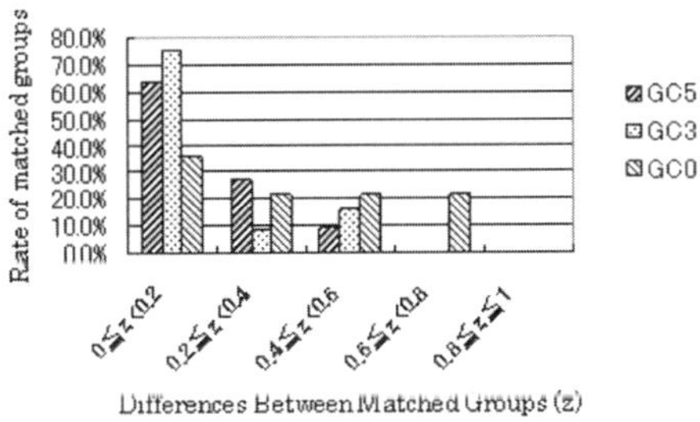

Fig. 3. Distribution of Differences between Group Contoribution Rates in Matched Groups

4 Discussion

The data suggests that a player's behavior can be controlled without changing the dominant strategy of the game. The contribution rates of GC0 and GC3 in this study were larger than 0, the theoretical prediction, and larger than the contribution rate of the benchmark treatment. Another and more noteworthy finding was the similarity in the contribution levels between GC3 and GC5. The players contributed at high rates which did not decrease by iterations. Why did the players contribute so generously in GC3? There are two factors that can encourage cooperative behavior-stakes and information. In GC0 the stakes were zero, but the players were informed of how many individuals of the opponent group selected the contribution strategy. With this information, the players could ascertain whether their group had won, lost, or tied with the opponent group in the contribution level at the end of each period. The contribution rate of GC0 was high compared with the benchmark result. This difference between the treatments demonstrates the significantly positive effect of the information on the performance of the opponent groups. Thus, part of the high level contribution of GC3 might have been encouraged by the information. Unlike the case with many real world problems, subjects in this study gained feasible advantages by attaining cooperative outcomes. The small group size of only two members in our experiments enabled the subjects to monitor each other. This may have helped them attain a high contribution rate. A larger group size, on the other hand, could dilute the effect of competition. We should also keep in mind that the condition of each member is homogeneous in this game. The difference in the marginal cost for the group project or in the marginal profit from the group project may weaken an individual's desire to contribute. These issues need to be studied further in future experiments.

References

1. Bornstein G. (1992) The Free-Rider Problem in Intergroup Conflicts Over Step-Level and Continuous Public Goods. Journal of Personality and Social Psychology 64, No. 4:597–606
2. Bornstein G., Ben-Yossef M. (1994) Cooperation and Single Group Social Dilemmas. Journal of Experimental Social Psychology 30:52–67.
3. Bornstein, G. Erev I., Goren H. (1994) The effect of repeated play in the IPG and IPD team games. Journal of Conflict Resolution 38, No.4:690–707.
4. Fehr, E., Gaechter, S. (2002) Cooperation and Punishment in Public Goods Experiments. American Economic Review 90:980–94.
5. Goren H. (2001) The Effect of Out-Group Competition on Individual Behavior and Out-group Perception in the Intergroup Prisoner's Dilemma (IPD) Game. Group Processes & Intergroup Relations Vol4(2):160–182.
6. Goren H., Bornstein G.(1999) Reciprocation and Learning in the Intergroup Prisoner's Dilemma game. In: D. Dudescu, I. Erev, and R. Zwick (eds) Games and human behavior: Essays in honor of Amnon Rapoport. Hillsdale, NJ: Erlbaum
7. Nalbantian, H. R., Schotter A. (1997) Productivity Under Group Incentives: An Experimental Study. American Economic Review Vol. 87, No. 3:314–341.

The Mixed Effect of Voluntary Revelation: Evidence from Threshold Public Goods Game Experiments *

Yasuyo Hamaguchi

Nagoya City University

1 Introduction

A number of experimental studies on public goods have shown that many people prefer to cooperate with others in the beginning, but this tendency diminishes as they interact with others in the course of the experiments (e.g. Andreoni [1]). Most experimental works on public goods games have so far tried to eliminate social interaction among subjects as much as possible. The purpose of such artificial procedures is to observe people's selfish motivation uncontaminated by social norms. However, this sort of approach might be misleading if one wants to understand how people successfully produce or maintain a public good in reality.

The main focus of this paper is the effect of voluntary revelation of people's action. Since the motivation of voluntary revelation must be affected by what kind of reward structure is given, two different kinds of threshold public goods experiments were conducted. In both games, subjects can get a benefit from the public good if their group can collect enough contributions to the public good. In one of the games, people get a fixed payoff from the public good regardless of the amount of contributions collected (as long as they achieve or exceed the threshold level). In the other game, the more contributions are collected, the more people can get from the public good. The difference between the two games is how the excess amounts of contributions are distributed among people. In spite of this difference, the two games have the same Nash predictions.

* This research was done as a part of the research project "Experimental Economics: A New Method of Teaching Economics and the Research on Its Impact on Society," which is supported and funded by the Ministry of Education, Culture, Sport, Science and Technology of Japan and Kyoto Sangyo University. The author thanks Sobei H. Oda for his generous support for this research and also thanks Yoshio Iida for his technical advice for programming experiments using z-Tree.

Ledyard [4] summarized experimental works in threshold public goods games and observed that adding a threshold level to the linear voluntary contribution mechanism has a positive effect on people's cooperative behavior. Cadsby and Maynes [2] conducted a threshold public goods game experiment and found that subjects converge toward a different equilibrium (efficient or inefficient) depending on what kind of social background they have. Marks and Croson [5] examined three rebate rules of threshold public goods games and discovered the rate of equilibrium conversion and the variance of contributions differ significantly among the rules. Our threshold public goods games are similar to the"No Rebate" and"Utilization Rebate" rules in Marks and Croson.

To see the effect of voluntary revelation, subjects were allowed to reveal their own contribution to other group members if they would like to reveal it. Although adding such a revelation stage does not change the equilibrium prediction, experimental results show that the effect of the revelation opportunity was significantly different in the two games. In particular, the results over the short term (10 periods) were totally opposite from those over the long term (50 periods).

2 Experimental Design

The two games employed are the same as the threshold public goods games used in Marks and Croson [5], except for the"money back guarantee rule." In their experiment, each individual contribution was refunded in the case the group could not achieve the threshold level of contributions, while the contribution was not refunded in my experiment. One game is called"No Rebate rule," and the other game is called"Utilization Rebate rule," following the terminology of Marks and Croson [5].

In the No Rebate rule, one group consists of five members. They are asked to contribute to a public good (g_i) from their individually given endowments (=5 tokens). If they can collect contributions more than or equal to a threshold level (=10 tokens), each person can receive a benefit from the public good (=4 tokens) regardless of how much they contributed. Therefore, the utility function of each person (U_i) in the No Rebate rule is derived as follows:

$$U_i = (5 - g_i) + 4 \quad if \sum_{i-1}^{5} g_i \geq 10,$$
$$U_i = 5 - g_i \quad otherwise.$$

The set of efficient Nash equlibria consists of any combination of $\{g_1, g_2, ..., g_5\}$ such that $\sum_{i-1}^{5} g_i = 10$ and $0 \leq g_i \leq 4$. On the other hand, the inefficient Nash equilibrium is that all individuals contribute nothing. The Pareto efficient outcome coincides with the efficient Nash equilibria.

The Utilization Rebate rule is different from the No Rebate rule only in how the excess amounts of contributions above the threshold are distributed among people. In the No Rebate rule, the excess amounts of contributions are wasted and no benefit is created from them. In the Utilization Rebate rule, the excess amounts of contributions are distributed among people equally. Therefore, the utility function of each individual is derived as follows:

$$U_i = (5 - g_i) + 4 + \frac{2(\sum_{i-1}^{5} g_i - 10)}{5} \quad if \sum_{i-1}^{5} g_i \geq 10,$$

$$U_i = 5 - g_i \quad otherwise.$$

The inefficient Nash equilibrium is that all individuals contribute nothing, which is the same as in the No Rebate rule. The set of efficient Nash equllibria is also the same as in the No Rebate rule. However, the Pareto efficient outcome in this rule is such that all individuals contribute all initial endowments.

Under the above reward structures, the experimental procedures consist of two stages as follows:

Stage 1 (Announcement Stage): Players announce to other people in their group whether they will show their individual contributions after stage 2.

Stage 2 (Contribution Stage): Upon observing the decision making of others in stage 1, players decide how many tokens to contribute to the public good.

In stage 2, subjects decide how much to contribute to the public good upon knowing who (indicated by an anonymous ID) wants to reveal their contri butions at the end of stage 2,. At the end of stage 2, only the decisions of people who decided in stage 1 to reveal their contributions are shown to the other group members. The decisions of people who decided not to reveal their contributions are not shown to anyone else. If the decision to reveal successfully conveys the signal"I want to cooperate with you," then adding such an announcement stage might lead people to converge toward an efficient equilibrium. Two sessions consisting of 50 periods were conducted. Five five-member groups made decisions repeatedly with the same partners over 50 periods in each session. Since there might be some ordering effect of the two rebate rules, the ordering of one session (Session 1) is No Rebate rule first and Utilization Rebate rule next, and the ordering in the other session (Session 2) is Utilization Rebate rule first and No Rebate rule next.

3 Results

Figure 1 shows group contribution levels of all groups in each treatment across periods. The top graphs of the figure are for No Rebate Rule and the bottom graphs are for Utilization Rebate rule. The obvious difference between the two rules is the fluctuation of group contribution levels. Group contributions in No Rebate rule fluctuate up and down in approximately the first half of the sessions and then remain stable at the threshold level (cooperative equilibrium) or the non cooperation level. Although some groups already successfully focused on the threshold level of contribution in the early periods, most other groups failed to produce the public good in early periods.

Some group contributions of those unsuccessful groups decayed to the zero contribution level eventually, and others recovered up to the threshold level of contribution. In No Rebate rule of both sessions, three groups converged toward the cooperative equilibrium, one group converged toward the non cooperative equilibrium, and the remaining group contributed some positive amounts but could not achieve the threshold level till the end.

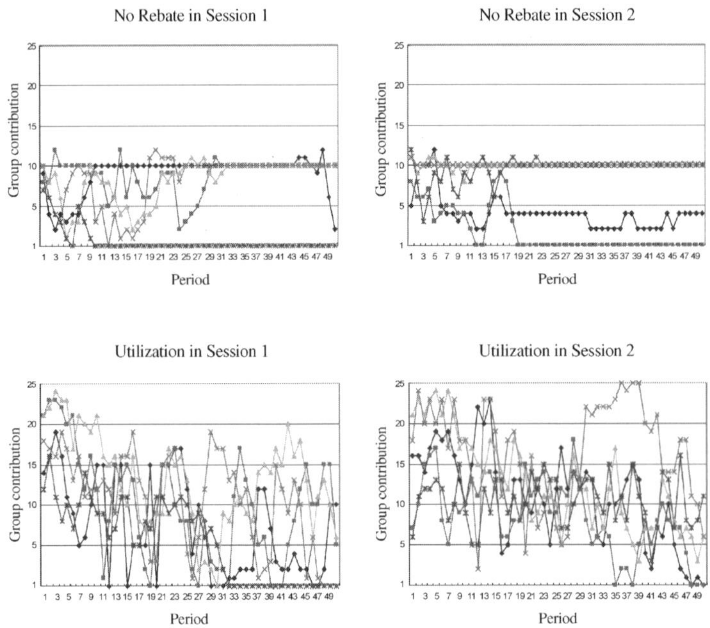

Fig. 1.

In Utilization Rebate rule, although a high contribution level is observed in the first 10-period interval, it was just a temporary phenomenon. Only one group converged toward the cooperative equilibrium in each session, two groups converged toward the non cooperative equilibrium, and the remaining two groups failed to produce the public good by contributing less than the threshold level.

These results imply that in spite of the lack of a payoff incentive for excess contribution, the No Rebate rule is a better payoff structure to maintain the threshold level contributions in the long term.

4 Conclusion

This study examined the effect of voluntary revelation of individual actions on people's cooperative behavior. One interesting finding of this study is that the content of pre-play communication matters in both a positive way and a negative way. The effect of pre-play communication depends on how long people interact with others and what kind of reward structure they are given. In the short term, under the Utilization Rebate rule, group contributions are higher than those in No Rebate rule. However, such a positive tendency does not last very long. Subjects tended to be less cooperative as the experiment proceeded and failed to produce the public good frequently in the later half of the sessions. In contrast, the results of No Rebate rule show that although few groups produced the public good in the early periods, more groups tended to successfully produce the public good in the later half of the sessions. In addition, most subjects clearly focused on the cooperative equilibrium with equal cost sharing. Those cooperative groups could maintain the threshold level of contribution fairly stably until the end of experiment. Therefore, No Rebate rule seems to be a better mechanism in the long run than Utilization Rebate rule. This switch phenomenon indicates that the positive effect of marginal per capita return in the Utilization Rebate rule may not be robustly strong against other psychological factors and it can be easily cancelled out by the effect of negative reciprocity.

References

1. Andreoni, J., (1995),"Cooperation in Public-Goods Experiments: Kindness or Confusion," *American Economic Review*, 85, 891-904.
2. Cadsby, C. B. and E. Maynes, (1998),"Gender and Free Riding in a Threshold Public Goods Game: Experimental Evidence," *Journal of Economic Behavior and Organization*, 34, 603-620.
3. Fischbacher, U., (1999),"z-Tree - Zurich Toolbox for Readymade Economic Experiments - Experimenter's Manual," Working Paper Nr. 21, Institute for Empirical Research in Economics, University of Zurich, 1999.
4. Ledyard, J., (1995),"Public goods: A Survey of Experimental Research," *Handbook of Experimental Economics*, ed. J. Kagel and A. Roth. Princeton, Princeton University Press, 111-194.
5. Marks, M. and R. Croson, (1998),"Alterative Rebate Rules in the Provision of a Threshold Public: An Experimental Investigation," *Journal of Public Economics*, 67, 195-220.

Trust and Reciprocity in Intergroup Relations: Differing Perspectives and Behaviors

Fei Song

Ryerson University

1 Research Background

In practice, a manager of a firm needs to decide on behalf of his/her firm whether to trust a competitor and accept its proposal to collaborate on the development of a new product. In situations like this, such an individual is viewed as the agent of the group and the individual ' s decision could have important implications for everyone in the firm. In many other situations, decisions to trust or reciprocate in inter-group exchange can also be made by groups themselves, such as committees, or management boards, who spend time examining an issue together in order to reach a group-level consensus decision via a collective process. These scenarios raise many interesting and important theoretical questions: Do individuals trust / reciprocate more (or less) when they are given the responsibility to act on behalf of their groups in contrast to when they represent solely themselves? Do groups trust / reciprocate more (or less) in contrast to their individual members? Is there relationship between the trust and reciprocity exhibited by individuals, group agents, and consensus committees?

The present research looks into these issues. First, I extend research in the area by looking at the effects of perspective taking in such interactions. Trust and reciprocity decisions are interactive in nature in strategic interactions and therefore require predictions about the counterpart ' s preferences and decisions. Such social judgments or predictions are particularly important but also difficult tasks of everyday social life. Existing research has shown pervasive cognitive errors that bias individual decision-making (e.g., Gilbert, Pinel, Wilson, Blumberg, & Wheatley [7]; Kahneman & Tversky [9]; Loewenstein, O ' Donoghue, & Rabin [10]). I am interested in whether individuals in trust and reciprocity exchanges have realistic perspectives about the exchange and their opponents ' intentions and behavior. To this end, I examine whether there is a disconnection between expectation and actual behavior and whether the disconnection is affected by the fact that the decision-making agent being an individual, a group agent, or a group. Along this line of inquiry, I also explore how trust expectations and the level of actual trust received influence the level of reciprocity.

Second, I investigate the influences of intra- and inter-group dynamics on inter-group trust and reciprocity. I investigate whether trust and reciprocity cognition and behavior are qualitatively different in inter-group versus inter-individual exchanges. I posit that, when members of a group interact with another who is outside of his/her group/organization, they interact not as individuals qua individuals, but according to organizational categories and group with which they belong to and identify with.

Third, I investigate whether dynamics of trust and reciprocity differ in various inter-group interactions where inter-group decisions are operationalized as 1) individuals making decisions for their groups as group agents in an autonomous and private manner without consultation with their group members, henceforth the autonomous group agent mechanism, and 2) group members making a consensus decision for their groups via a collective process, henceforth the consensus committee mechanism.

2 Methodological Overview

A widely employed game-theoretic experimental framework, the trust game (Berg, Dickhaut, & McCabe [2]), is adopted in the present research.[1] The trust game is played as follows. Players are randomly assigned as either trustors or reciprocators and given a certain amount of endowment at the beginning of the game. A trustor decides how much of their endowment to send to an anonymous counterpart, a reciprocator. All players are informed that each dollar a trustor sends would be tripled by the time it reaches the reciprocator. After receiving the money, the reciprocator then decides how to split his/her wealth, i.e., the sum of his/her initial endowment plus the tripled amount received, between him/herself and the trustor as an act of reciprocity. Trust in this game is operationalized by the amount sent by the trustor, and reciprocity is operationalized by the ratio between amount sent by the trustor and the amount sent back by the reciprocator.

I also explore a particular type of social judgment: perspective-taking. Specifically, I explore three types of perspective-taking in strategic exchanges. The first type is *intrapersonal, cross-situational perspective-taking*, i.e., what would *I* do in *that* situation, hence "*self-forecast*". The second type is *interpersonal, cross-situational perspective-taking*, i.e., what would *that person* do in *that* situation, hence "*other-forecast*". In terms of the operationalization of self- and other-forecasts, from reciprocators I collect cross-situational perspective-taking of "trust self-forecast". In the analogous fashion, I examine those of trustors, namely "reciprocity self-forecast" and "reciprocity other-forecast". Based on these two types of cross-situational perspective-taking, I

[1] For a comprehensive review of the trust game see Camerer [3].

further explore whether people judge themselves and others similarly or dissimilarly and whether they show an egocentric bias in their perspectives about the self and others. Lastly, I also measure *interpersonal, intra-situational perspective-taking*, i.e., what would *that person* do in *this situation*, hence "member-forecast". Specifically, in the group agent conditions I ask participants to give behavioral forecasts of what their group members would do if they were representing the group.

In sum, based on the experimental framework of the trust game, the present research examines the effects of two manipulated factors. The first is a between-person factor: the random assignment of participants to the trustor or the reciprocator role. The between-person random assignment of participants as trustors or reciprocators allows me to gather trust/reciprocity behavior and self-, other-, and group member-forecasts of such behavior. The second factor is a within-person design of two experimental conditions: the individual and the group agent / group consensus conditions. The within-person factor is an important feature of this study. It is adopted for the purpose of controlling for individual differences in trust/reciprocity preferences. Specifically, each participant, either as a trustor or a reciprocator, sequentially made two decisions, one as an individual agent, and the other as a group agent or consensus committee member.

3 General Discussion and Research Contributions

This research addressed a theoretically important and empirically meaningful issue of how trust and reciprocity are transformed from the individual to the group and organizational level, as well as the interplay between trust and reciprocity as well as between behavior and cognitions. Many key findings of this research suggest that the level of trust and reciprocity relationship, i.e., the inter-personal or inter-group level, can make a big difference on the level of trust and reciprocity both in behavior and in cognition. More remarkably, the specific decision-making mechanism in place in an inter-group interaction also has complex impacts on the level and the interplay of trust and reciprocity relations.

Theoretical Contributions First, in regard to trust and reciprocity exchange, this research extends the literature by showing that trust and reciprocity are influenced by many factors. This current research suggests that individual dispositions, situation-bounded cognitions, and institution/culture may independently and jointly influence trust and reciprocity. More importantly, trust and reciprocity are fundamentally distinct constructs. They are influenced by different motivations and psychological processes and are manifested differently in behavior.

The second theoretical contribution concerns the social contextual nature of trust and reciprocity exchange. This current research shows that trust and reciprocity at the group level are not readily inferable from the individual-level counterparts. In Study 1, in a minimal group paradigm fashion, participants were put into groups of three in an *ad hoc* manner and were then asked to make a trust/reciprocity decision as a group agent on behalf of their groups autonomously and privately. The results of this study revealed that autonomous group agents exhibited significantly less trust / reciprocity than they did as individuals in an analogous situation. More remarkably, the level of trust and reciprocity of group agents are positively correlated with their expectation of implicit group norm, which was measured by group agents' expectation of what their group members would do in their position. Moreover, this research also revealed that, even though both trust and reciprocity were considered socially desirable attributes as participants showed a "positive illusion" bias on both trust and reciprocity, people exhibited a lower level of trust and reciprocity toward another group when they took the role of a group agent. This is a remarkable result as a group agent and interacting with another group agent may reduce one's social preference, trust or reciprocity in this case, even if such social preferences are normatively appropriate. These group agent results yielded in Study 1, together with my other study (Song, Cadsby, & Morris [11]), are the first to show that the discontinuity effect not only exists in group behavior as extensively reported by Insko and his various colleagues (e.g., Insko, et al. [8]) but also exists in group agent behavior.

Third, I showed that the choice of the group decision-making mechanism can make a big difference on the trust / reciprocity decision. An important implication is emerging here: the extent and direction of the discrepancy between individual and group choices in regard to trust and reciprocity levels and possibly other social preferences in general may depend importantly on the precise details of the group decision-making mechanism, for example whether decisions are made consensually, by majority vote, or by a group leader or agent. This warrants further study.

Fourth, in regard to the relationships between perceived norm, expectation, and behavior, this research showed that other-underestimation bias can have a remarkable effect on autonomous group agent behavior. The fact that people often hold skeptical or negative views about other people may have served an cognitive rationale for people to lower their level of trust and reciprocity when they take the group agent role in order to "better" represent their groups. In other words, when group norms are merely implicit, people may infer such implicit group norm based on their expectation of other people and such a cognitive process may in turn change one's behavior even though the underlying structure of the situation remains the same. Paradoxically, even though participants showed a behavioral discontinuity in most inter-group interactions as autonomous agents, collective consensus, or informed agent, they did not expect such a behavioral discontinuity in either themselves or others. Thus, this research, as a first study of this kind, suggests that there also exists disconnection between cognition and behavior in strategic interactions as people don't have an accurate prediction of what they would do or others would do in inter-group interactions.

Methodological Contributions Laboratory experimentation is a valuable tool for this research program: it allows me to compare the experimental conditions under equivalent situations and thus holding constant other factors that might affect variables of interest. The experimental design employed in this research is innovative in that it combines both within- and between-person design in such a way that it maximizes the precision and the power of the analysis and at the same time minimizes possible cross-contamination of various kinds of variables. First, this methodological approach is particularly appropriate for studying the independent influence of personalities, attitudes, and situational factors, which have been long debated by social psychologists and behavioral economists. Second, I examined the relationship between visible (gender, ethnicity, and age) and latent individual differences (personality, social preferences, and expectations) and their impact on trust and reciprocity. Third, the combination of expectation and behavior data also allowed me to investigate the impact of expectation of trust and reciprocity from the counterpart on actual reciprocity / trust behavior exhibited by the actor.

Contributions to Practice The empirical findings of this research have important implications for practitioners in terms of the following dimensions. First, it is critical to effectively manage schema-related cognitions in intergroup relations, such as inter-departmental relations within companies or inter-firm relations in joint venturing, mergers-and-acquisitions, and other forms of firm-level strategic alliances. I advocate that much managerial effort be spent on developing effective and efficient interpersonal and work group relations; minimizing interpersonal or inter-group conflicts and negative group processes; enhancing the well-being of group members; and maximizing productivity.

Second, it is important to pay attention to how decisions are made at the group or firm level in strategic interactions as the specific decision-making process and mechanism may profoundly change the nature and the interplay of such interactions. The collective process of decision-making by a committee can in fact be a double-edged sword: it promotes cognitive diversity but it may also evoke responsibility alleviation or morality dampening at the same time. Managers should be acutely aware of such negative effects. It is also useful for managers to be aware that a work-unit or company agent, appointed to make a decision on behalf of the work-unit or the entire company, may make considerably more group-focused and less socially beneficial decisions than the same individual might make on his/her own behalf.

A third major practical implication highlights the importance to overcome the cognition-behavior disconnections. With modern organizations, there is much interest in fostering effective intra- and inter-organizational relationships. This study underscores the ideas that more effective managers will have greater perspective-taking abilities. As Weick [8] urged managers to "complicate yourself", this research suggests that effective managers are those who are able to see and understand interactions and relations from not only their own perspective, but their counterpart's as well. Specific advice derived from the findings of this study include to take into account situational forces and be more attentive to the factors that are important to the other party in order to maximize both the individual and joint gain of the interaction.

References

1. Barney, J. B., & Hansen, M. H. (1994). Trustworthiness as a source of competitive advantage. Strategic Management Journal, 15, 175-191.
2. Berg, J., Dickhaut, J., & McCabe, K. (1995). Trust, reciprocity, and social history. Games and Economic Behavior, 10, 122-142.
3. Camerer, C. F. (2003). Behavioral game theory: Experiments on strategic interaction. Princeton University Press.
4. Carnevale, P. J., & Pruitt, D. G. (1992). Negotiation and mediation. Annual review of psychology, 43, 531-582.
5. Child, J., & McGrath, R. (2001). Organizations unfettered: Organizational form in an information-intensive economy. Academy of Management Journal. 44, 1135-1149.
6. Freidman, R. A., & Gal, S. (1991). Managing arond role: Building groups in labor negotiaitons. Journal of Applied Behavioral Psychology, 27, 356-378.
7. Gilbert, D. T., Pinel, E. C., Wilson, T., Blumberg, S. J., & Wheatley, T. P. (1998). Immune neglect: A source of durability bias in affective forecasting. Journal of Personality and Social Psychology, 75, 617-638.
8. Insko, C. A., Pinkley, R. L., Hoyle, R. H., Dalton, B., Hong, G., Slim, R., Landry, P., Holton, B., Ruffin, P. F., & Thibaut, J. (1987). Individual group discontinuity: The role of intergroup contact. Journal of Experimental Social Psychology, 23, 250-267.
9. Kahneman, D. & Tversky, A. (1979). Prospect theory: An analysis of decision under risk. Econometrica, 47, 263-291.
10. Loewenstein, G., O'Donoghue, T. & Rabin, M. (2003). Projection bias in predicting future utility. Quarterly Journal of Economics, 118, 1209-1249.
11. Song, F., Cadsby, B., & Morris, T. (2005). Other-regarding behavior and behavioral forecasts: females versus males as individuals and as group representatives. International Journal of Conflict Management, in press.
12. Weick, K. 1979. The Social Psychology of Organizing (2nd Ed.). Reading, MA: Addison.

Frames and Games

Jordi Brandts[1] and Christiane Schwieren[2]

[1] Institut d'Anàlisi Econòmica (CSIC)
[2] Universitat Pompeu Fabra

1 Introduction

Economists are proud of being able to point to situations in which the fundamental forces identified by economic analysis work inexorably. One such case is the effect that supply reductions have on prices in the international oil market. One can safely predict that prices will increase and that any attempts by politicians and journalists to prevent the increase by presenting the situation in a particular light will not work. It is known, however, that things are not always that simple. Numerous studies have shown that behaviour often depends on the way in which logically equivalent choice situations and strategically equivalent situations are described or presented to people. Such so-called framing effects have been identified in a number of different contexts. Kuhberger [8] surveys some of the relevant literature.

The existence of framing effects poses important challenges to the scientific analysis of society. After all, social science is based on the idea that human behaviour can be captured and understood by simplified representations of things. If framing effects are pervasive, if every change in the circumstances surrounding social situations affects people's decisions substantially, the analysis of humans' social behaviour will be an extremely difficult task. At the same time the question arises why people are affected by the framing of situations.

In this paper we study experimentally some of the possible limits of framing effects. We do that in the context of different representations of dilemma games. Andreoni [1] compares behaviour in a public good (PG) and in a certain public bad (PB) game. He finds that subjects are more willing to cooperate in the PG case, even though the two situations are strategically equivalent. This result invites further investigation, because it is not directly consistent with some of the recent models of social preferences like those of Fehr and Schmidt [6], Bolton and Ockenfels [3] and Charness and Rabin [5]. The result also differs from most other results on framing effects in this kind of games. First, the strength of the effect is surprising. Two meta-analyses (Levin et al.

[9]; Kuhberger [8]) recently showed that PG/PB frames are not as effective in producing "framing-effects" as is the classical Asian disease situation of Tversky & Kahneman [11]. Both meta-analyses see the reason for this in the specific structure of "game-theoretic" (Kuhberger) or "goal-framing" (Levin) situations. Both choices are risky, and it is not clear, which one is riskier. Furthermore, Andreoni's effect goes in the opposite direction from what has been found in most studies of PG/PB frames. Usually, the negative (loss, PB) frame has been found to have a stronger impact on responses than the positive (gain, PG) frame. For example, in a classical study by Brewer & Kramer [4], subjects left more of the common resource in the commons frame than in the PG frame.

Our experiments are motivated by two specific hypotheses about the limits of framing effects. The first is that there is a kind of — in a loose sense — continuity in the relation between frames and behaviour. If this were the case, then very small changes in the way a situation is presented would have minor effects and only larger differences would lead to larger differences in behaviour. It would mean that not any small change would matter. The second hypothesis is that variations in the parameters that govern monetary incentives lead to similar changes — in direction and magnitude — under different framing conditions. This hypothesis is motivated by the general notion that perhaps models should not be expected to accurately predict levels of behaviour but should have the ability of accommodating observed comparative statics in the sense of the shifts in behaviour in response to parameter changes. In our experiments small and large changes in the presentation correspond to what – intuitively – are minor vs. more extensive changes in the wording used in the experimental instructions. With respect to the variations in the relevant parameters we follow the approach of Goeree, Holt and Laury [7], who have subjects make decisions for different PG situations, which vary in several dimensions.

2 Basic Experimental Design

A PG game with ten different parameter-combinations and three different frames is the basis for this research. The parameters used were taken from Goeree et al. (see Table 1). Their study changed parameters such that the "external" and "internal" return of contributions differed between the situations analyzed. The external return is defined as the return the investment has for others in the group, whereas the internal return is the value of the investment for oneself. In our experiment, parameters were varied within subjects, i.e., each subject had to take ten contribution-decisions.

Frames were varied between subjects. First, we compared two frames with a subtle linguistic difference, as had been used before successfully by Brewer et al. [4]. One group of participants played the game with a simple PG frame, the same used by Goeree et al., describing a situation where money could be

Table 1. Parameter structure of the ten decisions

Decision	1	2	3	4	5	6	7	8	9	10
Group size	4	2	4	4	2	4	2	2	4	2
Internal return	4	4	4	2	4	4	2	4	2	4
External return	2	4	6	2	6	4	6	2	6	12

"invested" in the public account or "kept" in the private account. In the PB frame subjects had to make a choice between "keeping" money in the public account or "investing" in the private account. Thus, in the PG situations, subjects were asked to do something good, whereas in the PB situations subjects could avoid doing something "bad". Our third frame differed in more aspects from the first two frames, but kept payoffs constant. It was designed following the PB frame Andreoni has used. Now the difference to a PG is not just linguistic. The frame describes the situa-tion such that when investing in the private account, some money is taken from each player in the group, while investing in the public account doesn't affect others.

144 students at Universitat Autonoma de Barcelona, from various faculties, participated voluntarily, for performance-based payment in an experiment on decision-making. In one session participated between 12 and 20 subjects. In each session, subjects took ten different decisions, but were exposed to only one frame. The order of the decisions was kept constant across participants. At the end of the experiment, one situation of the ten was randomly chosen and participants were paid according to their decision taken in this situation. In addition, they received a show-up fee of € 3.

3 Results

The main result of our experiment is that differences between frames are less important than the parameters of the decision task for determining contributions to a common good. However, one frame did have a stronger effect on contributions. Furthermore, the effect of the frames differed between parameter-constellations. The first figure shows the average contributions for each frame and each decision. To allow for a comparison with Goeree et al.'s original results in the same situations, we added their data to our figures. The figure shows that there are no big differences between the frames in most decisions, but that there are differences for all frames between the parameter-constellations. It also shows that the frames do not always influence contributions in the same direction.

With respect to the percentage of the endowment contributed in each frame, over all decisions, there is no significant difference for our three frames. While Goeree et al.'s result is very similar, Andreoni's data differ largerly from ours. In a later step we discuss whether this might be caused by the specific parameters he is using. But, first we take another look at contributions, but now

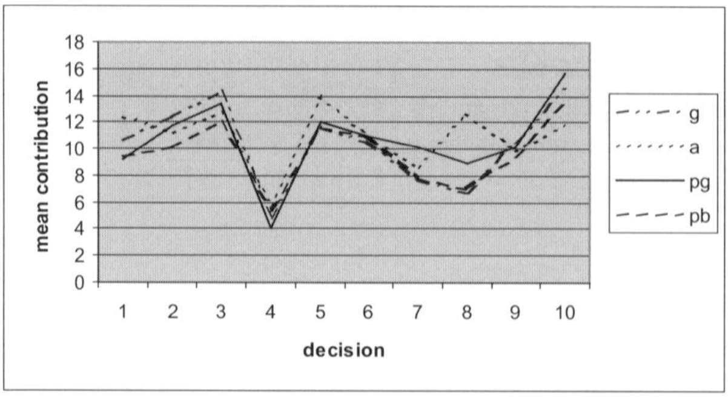

Fig. 1. Average token contributions per frame and decision (g = Goeree's data, a = our "andreoni"-frame).

looking only at the percentage of participants contributing zero to the PG. Doing this, differences between the frames become more pronounced. Specifically, our "Andreoni" frame differs significantly from our other two frames. It is nearly always highest with respect to the percentage of subjects contributing zero, and never lowest. However, in Andreoni's original study, the percentage of people contributing zero was even higher than in our version of his frame. A pos-sible explanation for this is that the parameters Andreoni uses are closest to our situation four with respect to internal and external return and num-ber of people in the social dilemma — and our situation four was exactly the one with the highest percentage of people contributing zero for all our frames. If we look additionally at the percentage of subjects contributing their full endowment in each decision and each frame, it becomes obvious that our frame "Andreoni" also has a very large number of subjects con-tributing fully. As it also has the largest percentage of subjects contributing zero in nearly all situations, average contribution in this frame looks very similar to the other frames, but this average stems from different behavior than in the other two frames.

Finally, one can look at whether there is a difference between treatments with respect to the percentage of subjects contributing more to the public account than to their private account. Differences exist between decision situations, but only for few situations between frames, and again, the "pb" and "pg" frame are very similar, whereas the "Andreoni" frame differs sometimes from the other two.

A statistical analysis (U-tests) shows no significant difference between the frames with respect to overall contributions (over all decisions). When looking at the percentage of full and zero contributions, differences between frames become significant, always the "Andreoni"-frame being the one that differs from the other two frames (significance levels always $< .005$). The linguistically different PG and PB frame differ with respect to the percentage of full contributions ($p < .005$). More people contribute fully in the PG than in the PB frame. With respect to zero contributions, both frames are equal. An ANOVA using "parameter structure" as independent factor and contribution as the dependent variable reveals that for each frame the type of decision is an important predictor of the amount contributed (significance level for PB and PB $< .01$ and for Andreoni and Overall $< .001$). Decisions are made independent for this analysis by subtracting contributions in each decision from contributions in decision one.

4 Discussion

Overall, we find only a weak effect of frames, which is in line with our hypotheses and with results from the meta-analyses cited above. We hardly find significant differences between the two only linguistically different frames, but we do find significant differences between the "more different" frame ("andreoni") and the two basic PG and PB frames. This confirms our hypothesis that stronger framing manipulations have stronger effects.

Our results further indicate that the direction of the framing effects seems to be influenced by the parameters chosen. This explains partially why Andreoni [1] contrary to most other studies on framing in PG/PB games finds lower contributions in his PB frame than in his PG frame. The parameter-constellation of our experiment which ist most similar to Andreoni's parameters results in the highest percentage of zero-contributions of all our constellations, and is close to the percentage of zero-contributions Andreoni reports. Another important aspect about the "andreoni" frame used in our experiment is that it leads to the largest variance in results. This frame has both the highest percentage of zero-contributions and the highest percentage of full contributions in most decision-situations. Unfortunately, Andreoni does not report on percentage of full contributions in his experiment.

Boettcher [2]concludes from his review of the existing literature on framing that "Relatively minor differences in experimental design appear to exaggerate or minimize the impact of prospect framing" (p. 355). Andreoni's study compared to our experiments is a nice example - if the "right" parameters are chosen, larger framing effects can be found, whereas the "wrong" parameters lead to no or very small framing effects. Furthermore, the effect of frames can go in both directions as the results with our "andreoni"-frame show: It leads both to more zero and more full contributions.

Our results are far from being conclusive. They confirm, that "goal frames" are more complicated than simple Asian-disease problems (Levin et al., [9]), because more than one aspect of the message can be manipulated, and because it is not obvious which option is the riskier one. Furthermore, there is room for differences in emotional intensity induced by different terminologies used. Andreoni's frame might enhance emotional intensity as opposed to the two other frames we use, as taking money from someone probably is emotionally more involving than just making different contributions.

One important problem of our design that could provide an alternative explanation for the results found is that each subject went through all ten decisions. This might have some demand-characteristics, inducing subjects to think carefully about the decision. Research by McElroy and Seta [10] has shown that subjects are far less susceptible to framing manipulations when they are prone to or asked to think analytically about their decision.

Our review of the literature showed that there exist some first attempts to characterize situations in which framing effects occur in PG/PB frames. However, there is no conclusive evidence yet and systematic research is lacking on what aspects of the frame and the parameters determine whether framing has an effect on subjects or not in this kind of situations. Our research wants to be a first step in this direction. Apart from providing some tentative results showing what factors might influence the effectiveness of framing manipulations, if underlines the necessity of further, more systematic research in this direction.

References

1. Andreoni, J (1995) Warm-Glow Versus Cold-Prickle: The Effects of Positive and Negative Framing on Cooperation in Experiments. The Quarterly Journal of Economics 110:1-21
2. Boettcher III WA (2004) The Prospects for Prospect Theory: An Empirical Evaluation of International Relations Applications of Framing and Loss Aversion. Political Psychology 25:331-362
3. Bolton GE, Ockenfels A (2000) ERC: A Theory of Equity, Reciprocity, and Competition. American Economic Review 90:166-193
4. Brewer MB, Kramer RM (1986) Choice behavior in social dilemmes: Effects of social identity, group size, and decision framing. Journal of Personality and Social Psychology 50:543-549
5. Charness G, Rabin M (2003) Understanding Social Preferences with Simple Tests. General Economics and Teaching 0303002, Economics Working Paper Archive at WUSTL.
6. Fehr E, Schmidt K (1999) A Theory of Fairness, Competition and Cooperation. Quarterly Journal of Economics 114:817-868
7. Goeree JK, Holt C, Laury S (2002) Private Costs and Public Benefits: Unraveling the Effects of Altruism and Noisy Behavior. Journal of Public Economics 83: 257-278
8. Kuehberger A (1998) The Influence of Framing on Risky Decisions: A Meta-analysis. Organizational Behavior and Human Decision Processes 75:23-55
9. Levin IP, Schneider S, Gaeth GJ (1998) All Frames Are Not Created Equal: A Typology and Critical Analysis of Framing Effects. Organizational Behavior and Human Decision Processes 76:149-188
10. McElroy T, Seta JJ (2003) Framing effects: An analytic-holistic perspective. Journal of Experimental Social Psychology 39:610-617

Combinatorial Auction Bandwidth Trading: An Experimental Study

Charis Kaskiris[1], Rajul Jain[2], Ram Rajagopal[3], and Pravin Varaiya[4]

[1] University of California, Berkeley
[2] University of California, Berkeley
[3] University of California, Berkeley
[4] University of California, Berkeley

1 Introduction

We study the interaction between internet service providers who lease bandwidth from owners of individual network links to form desired routes. Bandwidth markets were pioneered by Enron in the late 1990s by providing pooling points for switching and interconnecting. Williams Communications and RateExchange followed suit with their own markets [1]. Unfortunately, market dynamics, technical difficulties, and the collapse of Enron have brought these markets to an end. Two main factors hampered the implementation of these markets [2]. The first one has to do with the excessive time needed to disseminate new routing information, which is an artifact of the bilateral nature of contracting for routes. The second has to do with balance loading once carriers become multi-connected. These dynamic markets were setup to deal with the inefficient bilateral negotiations used in the industry. The inefficiency of bilateral contracting is due to the exposure effect from not being able to form attractive routes from individually leased links. This is a more general problem identified with markets of complementary goods/services. The combinatorial auction market mechanism has been proposed to alleviate the exposure effect [9]. Combinatorial auctions have been recently theoretically explored as a mechanism for bandwidth allocation [3, 7, 5]. Jain and Varaiya [5] propose a double-sided combinatorial auction mechanism for bandwidth trading whose implementation properties we explore using economic experiments in this paper.

In particular we compare the efficiency and bidding properties of the particular combinatorial auction setup in the presence of complementarities among the objects being allocated. Specifically, we conduct laboratory experiments allocating three links with private values and complementarities using the combinatorial auction format under different degrees of complementarity. In the benchmark case, every seller owns all types of links and every buyer has private values over all subsets of links. In alternative cases, sellers own two types of links with one type being owned by all sellers. Buyers have valuations over bundles of links but only over one singleton link.

The remainder of the paper proceeds as follows: section two presents the theoretical framework of the combinatorial seller bid double auction (c-SeBiDA); section three gives an overview of the experimental design. Section four reports the results of the experiments. We conclude with a discussion.

2 Theoretical Framework

The formal mechanism is described in Jain and Varaiya [5]. The mechanism is a sealed-bid, double-sided combinatorial auction format, with multiple buyers and sellers. The mechanism receives all the bids and matches buy with sell bids together to generate the maximal surplus. Uniform prices for each individual link are assigned on the basis of the highest winning sell bid price. The payments made by buyers to sellers are determined by the number of trunks at the determined uniform prices. The bid expression language also allows for different logical constraints applied on bids, namely XOR and OR constraints. These properties of the mechanism have not been experimentally tested yet. One further consideration is with regards to the total payment made by buyers. It is conceivable that the total amount owed may exceed the bid level, then that bid is rejected as a bigger surplus can be generated by removing it. Therefore the outcome of the auction respects individual rationality.

Strategic analysis of the mechanism is presented in Jain and Varaiya [5]. Under full information, Nash equilibrium exists for the combinatorial seller bid double auction, with everyone's strategy being truthful revelation, with the exception of the matched seller with the highest bid per link. In the complete information case the mechanism is always efficient, budget-balanced, ex-post individual rational and "almost" dominant strategy incentive compatible. In the incomplete information case, the resulting equilibrium allocation is always efficient, budget-balanced, ex-post individual rational, asymptotically efficient and Bayesian incentive compatible. Truth-telling is a dominant strategy for all the players but the player with the highest sell bid for each link. We explored the predictions of this theory in an experimental setup.

3 Experimental Setup

We conducted a combinatorial double-sided auction for a linear network with three links A, B and C. A linear network is totally unimodular, and it has been proved that a competitive equilibrium exists in such cases [4]. A valuation matrix for each player was generated for every round. The generation of valuation matrices for each player follows the suggestion in [8] and is described in [6]. For each seller, the cost valuation of a single trunk on each link was independently drawn from a uniform integer distribution between 5 and 15. For multiple trunks of a given link, the value was determined by the product of the number of trunks and the base value of the link.

For each buyer, a valuation for one unit of each of links A, B and C was drawn from a uniform integer distribution between 10 and 20. The valuation of a single trunk with link subsets (e.g. AB) was generated by adding the value of the two base items, and then adding a random integer between 0 and 5. Finally the single trunk, three link subset was determined by adding a uniform number on $\{0, 1, 2\}$ to the maximum of two link subsets. For each additional trunk, the value was the base value times the number of trunks minus a penalty. The penalty was 1 for a single item trunk, 2 if a double item trunk or 3 if all items were used. The idea of this valuation generation scheme was to model complementarity between goods from the buyers perspective.

Sellers are allowed to submit multiple bids, but each bid is only over a single link. Buyers may bid on combinations of links, but are restricted to have an equal number of trunks of each link for every bid. The bids are loose for buyers and sellers. In these particular experiments, seller bids are ORed and buyer bids are XORed. The objective of each bidder is to improve her endowment position through trading. Players are induced to perform well by rewarding changes from the initial endowment. At the end of all rounds conducted a fixed amount of money, \$500, is split among players in proportion to the total surplus generated, with negative balances being ignored.

The experiment consisted of a 3-hour experimental session which was conducted in August 2004 at the xLab facilities in the Haas Business School at the University of California, Berkeley. Eight subjects were recruited among graduate students in electrical engineering, information management and systems, and economics. Four buyers and four sellers were randomly assigned on each round. A new valuation for each player was generated every single round. For the first rounds the valuations were generated as described. In the last four rounds, an additional constraint on the valuation was incorporated: sellers were only assigned non-zero values for two links, and buyers had non-zero only for a single link (instead of 3) and multiple link combinations. Buyers were endowed with money at the beginning of a round. Sellers were endowed with three trunks on each link.

4 Experimental Results

We are interested on how the mechanism has worked under the two different market conditions with respects to market efficiency in allocating the links to the most efficient players. We also further interested in the strategies used by the players under the different market conditions and as they gained experience.

4.1 Overall Efficiency

In every round of each session has a predetermined total surplus based on the randomly assigned valuations. After each round, the surplus generated by the matched buyers and sellers is aggregated and the total surplus generated is calculated. Comparing the predetermined with the empirical surplus allows us to measure the percentage of total surplus generated per round per session which are presented in table 1.

Table 1. Efficiency Percentage Per Round for Each Session

Round	Session 1	Session 2
1	66.67	83.02
2	41.03	92.71
3	91.11	83.74
4	77.67	88.59

We observe that the average efficiency of the mechanism has been around 67% in the session 1 and about 87% in the session 2. The decrease in efficiency in round 2 of experiment 1 was mainly because a particular participant got confused and bid as a buyer when s/he was a seller. Even though we removed the bids submitted by the subject, given the small number of buyers and sellers in the experiment, it let to a decrease in overall efficiency. If we remove the efficiency result from that round we observe an average efficiency of 78%. From the results, we cannot conclude that more experience was leading to better efficiency results in either of the two sessions.

4.2 Bidder Behavior

We are further interested in the bidding behavior of the buyers and sellers in both sessions and during the different sessions. We are interested in the behavior of shading ones scores. Shading is defined as the percentage points of difference between the true valuation and the player's bid price. If it is positive, it means the players bid lower than their true valuation whereas when negative it means the players bid higher their true valuation. Shading is expressed in percentage point terms of the true valuation, e.g. 1 represents a bid 100% higher than the true valuation. Figure 1 presents the buyer and seller shading behavior across experiments and rounds.

We observe a dispersion in buyer bidding strategies. The distributions of bidding behavior in session 1 shows that the median underbidding by about 20% which varied across rounds. What we also observe is the existence of random strategies where bidders were both overbidding and underbidding substantially. In session 2 we observe similar behavior even though what we observe that as the rounds progressed the median shading behavior decreased and the dispersion of the shading behavior decreased.

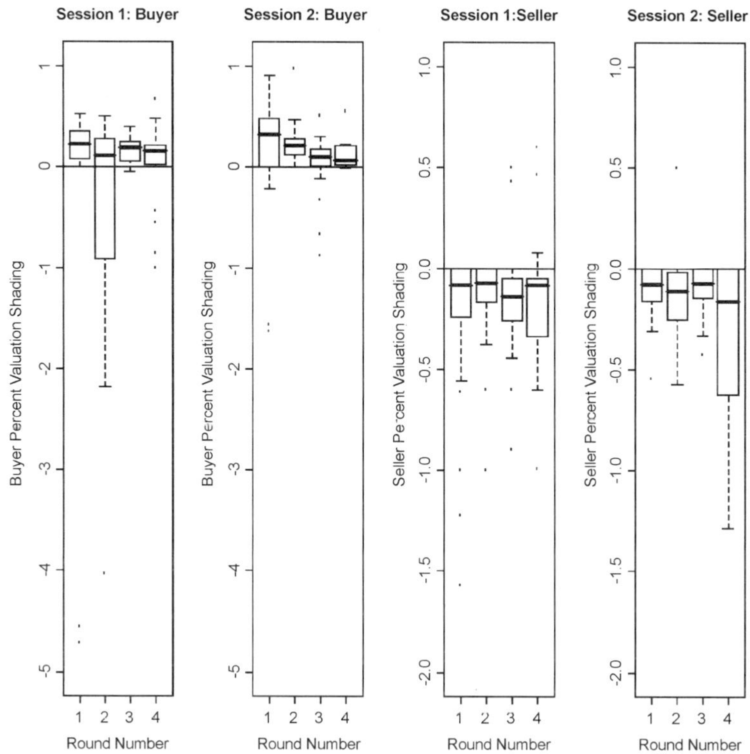

Fig. 1. (a) Boxplots of Buyer Shading Behavior Across Different Rounds in Session 1 (b) Boxplots of Buyer Shading Behavior Across Different Rounds in Session 2. (c) Boxplots of Seller Shading Behavior Across Different Rounds in Session 1 (d) Boxplots of Seller Shading Behavior Across Different Rounds in Session 2.

We observe a similar dispersion in seller bidding strategies. The distributions of bidding behavior in session 1 shows that the median overbidding by about 10% which varied across rounds. Again we observe patterns of random bidding behavior. The overbidding behavior is rather non-rational as if those bidders were the highest matched, then they wouldn't be able to recoup their costs. In session 2 we observe similar behavior as in session 1 without and distinct pattern across rounds. Even though the theory predicts that the sellers will be the ones who do not bid truthfully, their bidding is closer to truthful than the buyer bidding behavior.

A certain sense of strategic play is captured in the plots, as buyers tend to bid below their valuations when the valuation is high and sellers above their valuations when the valuation is low. Most likely they base their guess in the knowledge of the distributions used to generate valuations, given at the beginning of each set of rounds. This kind of behavior suggests that with eight players, the auction does not induce truthful behavior. But as the number of player increases, the strategy of guessing when valuation is low or high has a smaller probability of being correct. In fact, as the number of players goes to infinity this probability is zero.

5 Discussion

We performed an experimental evaluation of a combinatorial auction mechanism, using a real time auction setting to obtain empirical benchmarks with respect to the theoretically predicted results. The mechanism exhibits an efficiency loss in the finite player case, confirming the indications from the theory developed. Also, there are some indications that the mechanism does not clearly induce truth-revelation among bidders. As a future direction, an evaluation of the mechanism under a larger number of players and also under a sequential game setting will be performed. Another direction is to expand the number of links and at the same time allowing experimental subjects to define bids for source-destination pairings and have the engine deal with formulating the surplus-maximizing combination of links to do so and at what cumulative price.

References

1. Borthick, S. (2001). "Bandwidth Trading: What's in it for you?," Business Communications Review, pp. 26-30.
2. Ferreira, P., Mindel, J., McKnight, L. (2003). "Why have bandwidth trading markets not matured? Analysis of technological and market issues," International Journal of Technology, Policy and Management, 3(2), pp. 142 - 160.
3. Huang, J., Berry, R. and Honig, M. (2004). Auction-Based Spectrum Sharing. 42nd Annual Allerton Conference on Communication, Control and Computing.
4. Jain, R. and Varaiya, P. (2004a). "Combinatorial Bandwidth Exchange: Mechanism Design and Analysis," Communications in Information and Sciences, 4(3):305-24.
5. Jain, R. and Varaiya, P. (2004b). "An Efficient Incentive-Compatible Combinatorial Market Mechanism" Proc. of Allerton Conference.
6. Kaskiris, C., Jain, R., Rajagopal, R., and Varaiya, P. (2004). "Combinatorial Auction Design for Bandwidth Trading An Experimental Study," International Conference Experiments in the Economic Sciences, Okayama and Kyoto, Japan.
7. Maille, P. and Tuffin B. (2004). Multi-Bid Auctions for Bandwidth Allocation in Communication Networks, INFOCOM.
8. Morgan, J. (2002). "Combinatorial Auctions in the Information Age: An Experimental Study," M. Baye, Advances in Applied Microeconomics. JAI Press.
9. Porter, D. Rassenti, S., Roopnarinen, A. and Smith, V. (2003). "Combinatorial Auction Design." Proceedings of the National Academy of Sciences, 100(19), pp. 11153-57.

A Note on Peer Enforcement by Selective Exclusion: An Extended Abstract

Jana Vyrastekova[1] and Daan van Soest[2]

[1] Tilburg University
[2] Tilburg University

1 Introduction

In economic experiments, spontaneous emergence of cooperation in social dilemmas has been shown to arise if individuals can impose pecuniary sanctions on others. Reciprocal individuals are willing to punish free riders even if they themselves incur costs when doing so (Ostrom et al. [10], Fehr and Gächter [3]). Although such direct pecuniary punishments occur in the real world (Cordell and McKean [2]), ordinary citizens usually do not have the right to destroy another person's property, nor do they have the authority to impose fines. What citizens can do, however, is to cease interaction with individuals who free ride in the social dilemma situation, and refuse to cooperate with them in *other* social or economic circumstances in which they meet. Indeed, our everyday interactions are embedded in a system of interpersonal relations (Granovetter [6]) that require cooperation by two or more individuals (cf. Bowles and Gintis [1]). For example, Japanese villagers, Irish fishermen, and inhabitants of the Solomon Islands have in common that they cut contact with fellow villagers who free ride with respect to fishing, thus denying them the benefits of cooperation in other economic activities (McKean [9], Taylor [13], and Hviding and Baines [7]).

This note summarizes the main conclusions of van Soest and Vyrastekova [12] with respect to whether embedding a social dilemma in a wider economic context can explain the spontaneous emergence of cooperation. We develop an economic experiment in which subjects play a finitely repeated Common Pool Resource (CPR) game –which captures the social dilemma situation– and they also interact in a "gift-giving game " which requires bilateral cooperation. The former game is set up in line with the work of Ostrom et al. [10]; in the latter, subjects can send gifts to each other, where the costs of giving them are smaller than the financial benefits obtained by the recipient. That means that in the latter game, it is socially optimal for all subjects to give gifts to all co-players, although it is privately optimal to free ride and not give gifts to others. We investigate whether thus embedding the social dilemma

–as captured by the CPR game– in a wider economic environment sheds light on the emergence of cooperation. We hypothesize that linking the two games allows subjects to unilaterally cease cooperation in the gift-giving game in order to discipline the behavior of others in the CPR game. We will refer to this type of sanctioning as the selective exclusion mechanism.[3]

The design of our experiment is as follows. Subjects participate in a game repeated for 25 periods, with a stage game consisting of the CPR game and then the gift-giving game (for details see van Soest and Vyrastekova [12]). This repeated game has only one subgame perfect Nash equilibrium: rational money-maximizing individuals always overharvest the CPR and never give a gift to any other individual. This prediction is independent of whether individuals interact with the same group of individuals in both constituent games, or not. We compare behavior of subjects who interact in the same group in both games (Linked treatment), and subjects who interact in two disjoint groups (Unlinked treatment) in order to assess the viability of the selective exclusion mechanism and its effect on the efficiency of CPR use.

Per treatment we collected experimental data on 8 independent groups consisting of five group members, and hence a total of 80 students participated. The experiments were computerized, using z-Tree (Fischbacher [4]) and subjects were not allowed to communicate otherwise than via the computers. All decisions in the experiment were formulated in a neutral language. During the experiment, the history of each of the two games (the extraction effort in the CPR game and the number of gifts given by all participants in the group) was available on the computer screen, but we did not instruct subjects explicitly to use this information across games and rounds in any way. The experiment lasted about 2 hours, and participants earned on average 19.30 Euro (including 5 Euro participation fee).

2 Summary of Treatment Effects

Figures 1 and 2 present the average group data with respect to CPR extraction effort and number of gifts given, respectively, as well as the average aggregate group extraction effort from a related experiment by Vyrastekova and van Soest [14], where the unregulated CPR game was played in isolation (i.e., not tied to another game). Clearly, the aggregate extraction effort in the Linked treatment is closer to the socially optimal level ($X^* = 30$) than in the Unlinked treatments as well as in the unregulated CPR game, where it is close to the subgame perfect Nash equilibrium level ($X^{NE} = 50$). A 2-sided Mann-Whitney U-test rejects the hypothesis of equal group extraction effort levels

[3] Note that selective exclusion is with respect to (voluntary) cooperation in the alternative economic activity; it does not refer to denying individuals the right of access to the common pool resource, as this is very often not legal/feasible in practice (see McCarthy et al. [8]).

in the Linked and Unlinked treatments. The average number of gifts provided per group is far above the Nash equilibrium prediction of zero gifts in all four treatments. Remarkably, the number of gifts given within a group does not depend on whether the CPR game and the gift-giving game are linked or not.

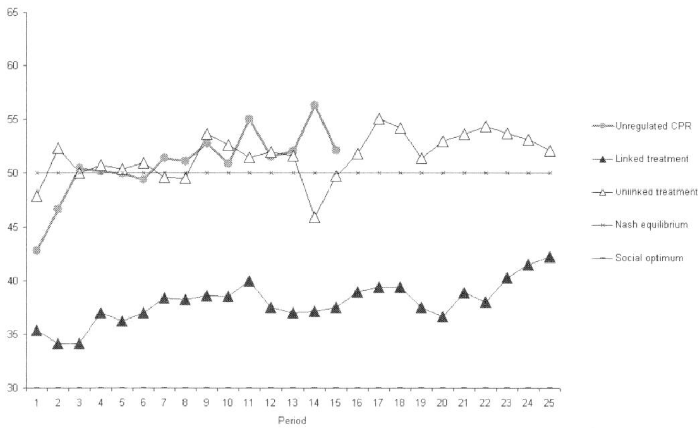

Fig. 1. Average group extraction effort in the CPR game

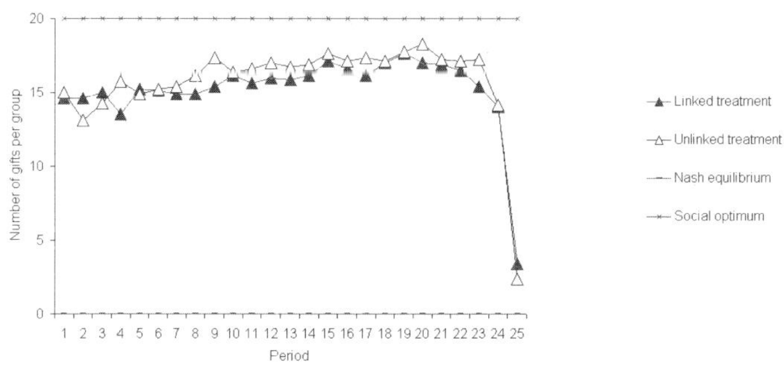

Fig. 2. Average number of gifts per group in the gift-giving game

All these observations are in conflict with the standard predictions of money-maximization and subgame perfect Nash equilibrium, but in line with the potential influence of reciprocal individuals on how the repeated game is played. This suggestion is supported by the analysis of the decisions made in

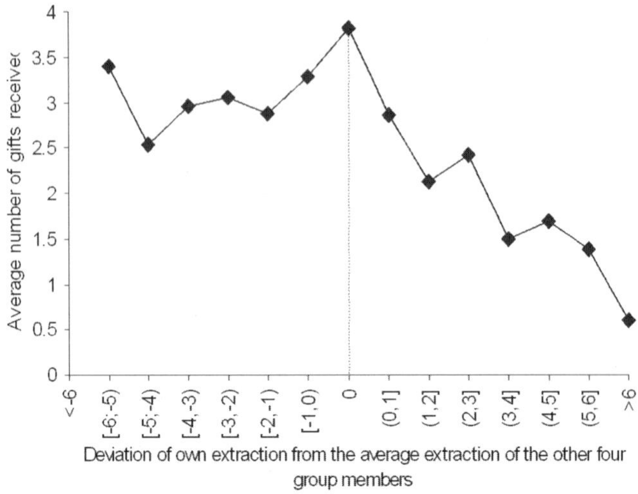

Fig. 3. Reciprocity across the two games in the Linked treatments

the Linked treatment, where the number of gifts given is found to depend on the extraction effort level chosen even in the first period: a subject is nearly twice as likely to receive a gift if he/she did not choose higher extraction effort in the CPR than his/her group did on average (p=0.000 according to a two-sided Mann-Whitney U-test).

Finally, evidence that decisions in the two games are interconnected in the Linked treatment is obtained when plotting the relationship between a subject's deviation in extraction effort from the group's average (i.e., $x_{i,t} - \frac{1}{N-1}\sum_{j\neq i} x_{j,t}$) on the horizontal axis, and the average number of gifts received on the vertical axis (averaged over all 25 rounds of the game); see figure 3. This figure suggests that putting in either more or less effort into CPR extraction than the group's average is correlated with fewer gifts received. Indeed, for extraction levels above the group average ($x_{i,t} > \frac{1}{N-1}\sum_{j\neq i} x_{j,t}$), the Spearman rank-based correlation coefficient for gifts received equals -0.886 ($p = 0.019$). However, for extraction levels below the group average ($x_{i,t} < \frac{1}{N-1}\sum_{j\neq i} x_{j,t}$), the correlation equals $-0.029(p = 0.957)$. Therefore, we conclude that in the Linked treatment, freeriding in the CPR game in the form of high extraction effort (as compared to the group's average) is correlated with fewer gifts received.

3 Conclusions

The group data summarized above suggest that embedding a social dilemma situation in a wider economic context (by adding the gift-giving game) gives rise to more cooperation in the social dilemma situation than predicted by economic theory. Moreover, whereas the pecuniary punishment mechanism results in a decrease in net efficiency (because of the 'deadweight loss' associated with the costs of imposing sanctions; see Ostrom et al. [10] and Ostrom et al. [11]: 176), the selective exclusion mechanism uncovered here results in a pure efficiency gain. Whereas aggregate extraction effort in the CPR game is closer to social optimum level in the Linked treatment than in the Unlinked treatment, the total number of gifts provided is identical. Linking the two games thus results in an unambiguous increase in the subjects' earnings: subjects in the Linked treatments earned significantly more than those in the Unlinked treatments (p=0.000).[4] Thus, our experiments suggest that strengthening community ties gives rise to powerful pro-social incentives with respect to cooperation in social dilemma situations and hence improves community welfare.

References

1. Bowles, S. and Gintis, H. (2002), "Social Capital and Community Governance", Economic Journal 112, 416-436.
2. Cordell, J.C. and McKean, M.A. (1992), "Sea Tenure in Bahia, Brazil", in D.W. Bromley and D. Feeny (eds.), Making the Commons work, San Francisco: ICS Press, 183-205.
3. Fehr, E. and Gächter, S. (2000), "Cooperation and Punishment in Public Goods Experiments", American Economic Review 90, 980-994.
4. Fischbacher, U. (1999), "z-Tree. Toolbox for Readymade Economic Experiments", IEW Working paper 21, University of Zurich.
5. Fischbacher, U., Gächter, S. and Fehr, E. (2001), "Are People Conditionally Cooperative? Evidence from a Public Goods Experiment", Economics Letters 71(3), 397-404.
6. Granovetter, M. (1985), "Economic Action and Social Structure: The Problem of Embeddedness", American Journal of Sociology 91, 481-510.
7. Hviding, E. and Baines, G.B.K. (1994), "Community-Based Fisheries Management, Tradition and Challenges of Development in Marovo, Solomon Islands", Development and Change 25, 13-39.
8. McCarthy, N., Sadoulet, E. and De Janvry, A. (2001), "Common Pool Resource Appropriation under Costly Cooperation", Journal of Environmental Economics and Management 42, 297-309.

[4] More detailed analysis of individual decision-making is found in van Soest and Vyrastekova [12], providing additional support for the selective exclusion mechanism.

9. McKean, M.A. (1992), "Management of Traditional Common Lands (Iriaichi) in Japan", in D.W. Bromley and D. Feeny (eds.), Making the Commons work, San Francisco: ICS Press, 66-98.
10. Ostrom, E., Gardner, R. and Walker, J. (1992), "Covenants With and Without Sword: Self-Governance Is Possible", American Political Science Review 86, 404-417.
11. Ostrom, E., Gardner, R. and Walker, J. (1994), Rules, Games and Common-Pool Resources, Ann Arbor: The University of Michigan Press.
12. van Soest, D.P. and Vyrastekova, J. (2004), "Economic Ties and Social Dilemmas", CentER Discussion Paper 2004-55, Tilburg University, Tilburg.
13. Taylor, L. (1987), "The River would Run Red with Blood: Community and Common Property in Irish Fishing Settlement", in B.J. McCay and J.M. Acheson (eds.), The Question of the Commons: The Culture and Ecology of Communal Resources, Tucson: University of Arizona Press, 290-307
14. Vyrastekova, J. and van Soest, D.P. (2003), "Centralized Common Pool Management and Local Community Participation", Land Economics 79(4), 500-514.
15. Vyrastekova, J. and van Soest, D.P. (2005), "Peer Enforcement in Common Pool Resource Experiments: Transfers, Rewards and Sanctions", Tilburg: Tilburg University, mimeo.

Part V

Performance of the System

Recycling of Durable Goods: Modeling and Experiments

Nariaki Nishino[1], Hiroki Nakayama[2], Sobei H. Oda[3], and Kanji Ueda[4]

[1] The University of Tokyo
[2] The University of Tokyo
[3] Kyoto Sangyo University
[4] The University of Tokyo

1 Introduction

This paper describes how durable goods can be recycled.

In general, a producer of recyclable goods faces an optimal pricing problem that is considerably complicated and difficult to solve mathematically because recycling is closely related to product durability. A producer's decisions related to durability might directly affect recycling performance. For example, the producer can not easily find a profit-maximizing price in such circumstances that the producer must set numerous parameters: new unit price, used unit price, recycled unit price, durability of new units, and durability of recycled units. Even if an optimal solution were derived, its state sensitively depends on a combination of production costs and recycling costs. Accordingly, such problems become further complicated.

Many studies have specifically addressed durable goods. For example, Swan considered questions of optimal durability [7]. Coase argued a problem of time inconsistency [2]. Recently, many studies have considered more realistic models of durable goods: the role of leasing [5], monopolized aftermarkets [1], and upgrade processes [3]. On the other hand, Reynolds conducted experiments with regard to durable goods theory and explained deviations from equilibrium using a version of bounded rationality [8]. However, studies that consider recycling problems as durable goods' problems are very few.

We have produced a model in which a monopolist produces new units of a durable good and sells them at price P, while collecting (purchasing) some used units it produced in the previous period at price Q to sell as recycled goods at price R. Mathematical analysis shows that the set of prices $(P,\ Q,\ R)$ that maximizes the monopolist's profit changes in a complicated fashion according to available technology and consumer preferences. In addition, we undertook a series of economic experiments to verify that people behave as the model suggests.

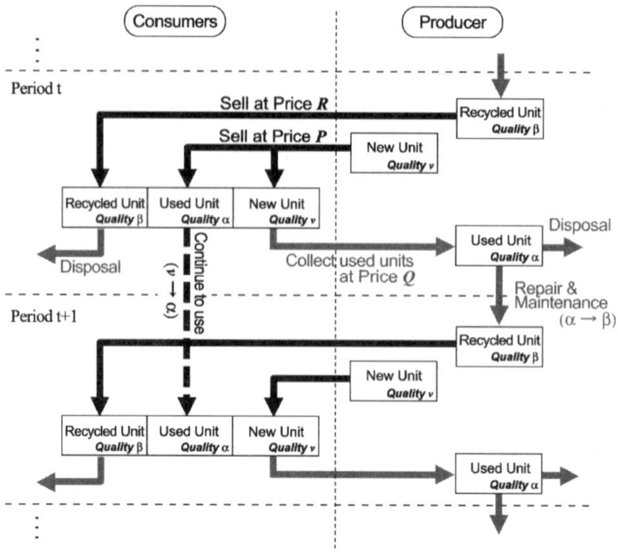

Fig. 1. Overview of the durable goods recycling model

2 The Model

A monopolistic producer, consumers and dismantlers exist in the market. An overview of our model is shown in Fig. 1. An explanation of each is as follows:

Durable goods

The good is useful for two periods, but its quality decreases from v (the quality of a new unit) to α (the quality of a used unit) after one period of usage. The producer can collect units to produce a recycled unit whose quality is β $(\alpha < \beta < v)$.

Consumers

Consumers demand a maximum of one unit at any date. A consumer of type θ who consumes a good with quality q obtains utility $u(\theta, q) = \theta q$, whereas a consumer purchasing a good at price p obtains surplus $V(\theta, q, p) = \theta q - p$. We assume that θ is distributed uniformly in $[0, 1]$. If a consumer purchases a new unit, the surplus is $\theta v - P$. If a consumer purchases a recycled unit, the surplus is $\theta\beta - R$. On the other hand, if a consumer purchases neither a new unit nor a recycled unit and uses the used unit that was purchased in a preceding period, then the payment is zero. In such a case, the surplus is calculated as $\theta\alpha - 0$.

Producer

The producer incurs a cost $c(v, \alpha)$ in producing a new unit and incurs a cost $d(\beta, \alpha)$ in producing a recycled unit. The producer has market power and chooses prices P, Q, R to maximize profit.

Dismantlers

Dismantlers always accept used-up units at price S from the producer. For simplicity, we assume that S is an exogenous parameter [5].

3 Mathematical Analysis

3.1 Formulation

We consider the long-run equilibrium in a steady state. In equilibrium, a consumer chooses one of the following consumption patterns according to each unit's utility parameters, price, and quality:

Pattern 1. a consumer purchases a new unit and sells the used unit in every period,
Pattern 2. a consumer purchases a new unit and continues to use it (purchases a new unit every two periods),
Pattern 3. a consumer purchases a recycled unit in every period,
Pattern 4. a consumer purchases no unit in any period.

We assume that x, y and z respectively represent quantities of Patterns 1, 2 and 3. Therefore, the producer's profit Π is represented as

$$\Pi = (P - c)(x + \frac{y}{2}) + (R - d)z - Qx, \tag{1}$$

where some constraints exist for prices and quantities (details of formulation are shown in [6]). In equilibrium, the producer chooses price P, Q, R to maximizc profit Π.

3.2 Equilibrium

Solving the maximization problem of eq. (1), the equilibrium is obtained. Although we omit the details here, equilibria are summarized in Fig. 2. From that figure, we conclude that the sales of recycled units depend on the conditions of cost c and d. Therefore, in a certain circumstance, even if the producer wants to disseminate recycled units into society, recycled units are not sold and are eliminated from the market. For example, no consumers purchase recycled units in Regions E and F, where recycling cost d is relatively large.

[5] The producer only collects used-up units at price S and enlists dismantlers to dispose of them at price S. We assume that the producer incurs no costs that are related to this disposal process. Therefore, the producer obtains no profit or loss in this process: the producer gets S and pays S.

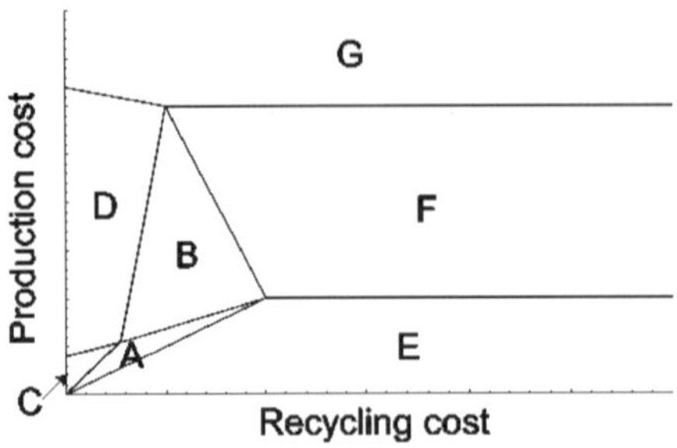

Region	Pattern 1	Pattern 2	Pattern 3
A	x_A	y_A	z_A $(< x_A)$
B	x_B	y_B	z_B $(= x_B)$
C	x_C	0	z_C $(< x_C)$
D	x_D	0	z_D $(= x_D)$
E	x_E	y_E	0
F	0	y_F	0
G	0	0	0

x_i, y_i and z_i $(i = A..F)$ respectively indicate the numbers of consumers who follow Patterns 1, 2 and 3 in the different regions.

Fig. 2. Equilibrium segmentation

4 Experiments

4.1 Experimental Design

The market comprises 100 consumers and a monopolistic producer. A subject plays the role of the producer, while each computerized agent plays the role of a consumer. Using this arrangement, we are able to isolate the effects of other subjects who play the role of consumers. We can thereby infer the manner in which a result is affected by behavior of particular subjects. Especially, the behavior of the producer, who makes the market, is crucial. The decision-making process of each is the following.

Producer

The producer determines prices P, Q, and R. If the producer sets prices, then the sales and the profit are calculated and are shown on the display.

Consumers

Each consumer calculates their own utility and makes a decision considering that utility. In other words, each agent selects one of four consumption patterns (Patterns 1–4) to maximize the utility. We assume that a preference parameter θ is distributed uniformly on $[0, 100]$.

4.2 Experimental Settings

As shown in Table 1, we conduct six kinds of experiments with respect to production cost c and recycling cost d. Durable goods quality is assumed as $v = 1.0$, $\alpha = 0.6$, and $\beta = 0.7$.

Experiments were conducted at the Kyoto Sangyo University Experimental Economics Laboratory (KEEL). Subjects were recruited from among undergraduate students at Kyoto Sangyo University. The number of subjects in each treatment is shown in Table 2. Subjects were rewarded according to their total profit, calculated as 0.01 yen per one point in the experiments.

Table 1. Treatments and costs

	Region	c	d
Treatment 1	Region A	26	10
Treatment 2	Region B	40	10
Treatment 3	Region C	18	4
Treatment 4	Region D	28	4
Treatment 5	Region E	26	18
Treatment 6	Region F	42	24

Table 2. Number of subjects

	Number of subjects
Treatment 1	9
Treatment 2	8
Treatment 3	8
Treatment 4	8
Treatment 5	7
Treatment 6	8

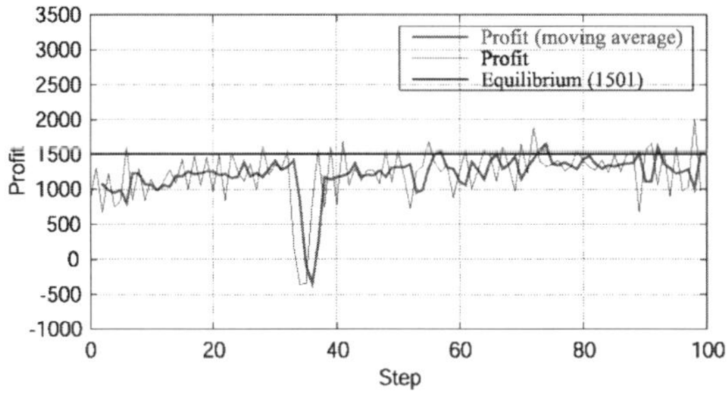

Fig. 3. An example of results (average producer's profit in Region A)

4.3 Results

Figure 3 shows the average producer's profit in Region A. We can infer that subjects realize the large profit around the equilibrium. However, most subjects do not select equilibrium prices because many near-optimal solutions exist in this profit-maximizing problem. Many subjects actually fall into near-optimal solutions.

This study does not specifically address whether a subject attains the equilibrium value; it instead shows how a recycling society can be formed. In other words, we examine whether recycled units are sold easily or not. Table 3 portrays sales of recycled units in each treatment. The table relates that, for treatments where the recycling cost is medium or large, the sales are not high regardless of the production cost. In contrast, for treatments where the recycling cost is small, the production cost should also be small to realize good sales of recycled units.

Results imply that recycling technology that lowers the recycling cost is more important than production technology in the case where the recycling cost is high. In other words, even if the production cost is high, the producer should lower the recycling cost. A consequent implication is that production technology that lowers the production cost is important to realize diffusion of recycled units in the case where the recycling cost is small.

Table 3. Recycled unit sales of respective treatments

		Recycling cost d		
		small	medium	large
Production cost c	small	Treatment 3 (C) Good sales		
	Medium	Treatment 4 (D) No sales	Treatment 1 (A) No sales	Treatment 5 (E) No sales
	large		Treatment 2 (B) Some sales	Treatment 6 (F) No sales

5 Concluding Remarks

This paper addresses the problem of durable goods recycling.

Economic agents such as producers or firms must make decisions in circumstances where the optimal pricing problem of durable goods is difficult to solve mathematically. Furthermore, the problem becomes more complicated when a producer considers recycling of durable goods.

We derive the equilibrium in the recycling market of durable goods, and conduct appropriate experiments. Those experiments demonstrated that subjects were unable to attain the equilibrium, but were able to attain a near optimal solution. Results engendered our conclusion that when both production cost and recycling cost are large, the recycled goods are not likely to be sold. Therefore, it is implied that the producer should assign priority to recycling-cost reduction.

Our analyses are generally useful for understanding decision-making. Additionally, they are applicable to problems involving institution and technology.

References

1. D. Carlton and M. Waldman (2001) "Competition, Monopoly, and Aftermarkets", NBER Working Paper series, no. 8086, National Bureau of Economic Research.
2. R. H. Coase (1972) "Durability and Monopoly", Journal of Law and Economics, Vol. 15, pp. 143-332.
3. D. Fudenberg and J. Tirole (1998) "Upgrades, Trade-ins, and Buybacks", RAND Journal of Economics, Vol. 29, No. 2, pp. 235-258.
4. I. Hendel and A. Lizzeri (1999) "Interfering with Secondary Markets", RAND Journal of Economics, Vol. 30, No. 1, pp. 1-21.
5. I. Hendel and A. Lizzeri (2002) "The Role of Leasing Under Adverse Selection", Journal of Political Economy, Feb., Vol. 110, No. 1, pp. 113-43.
6. N. Nishino, H. Nakayama, S. H. Oda, and K. Ueda (2004) "Recycling of Durable Goods: Modeling and Experiments", Proc. of International Conference Experiments in Economic Sciences: New Approaches to Solving Real-world Problems, pp. 521-536.
7. P. L. Swan (1970) "Durability of Consumption Goods", American Economic Review, Vol. 6, pp. 884-894.
8. S. S. Reynolds (2000) "Durable-goods monopoly: laboratory market and bargaining experiments", RAND Journal of Economics, Vol. 31, No. 2, pp. 375-394.

Decision Making in Artifactual Systems With Bounded Rationality

Tomomi Kito[1], Nobutada Fujii[2], and Kanji Ueda[3]

[1] The University of Tokyo
[2] The University of Tokyo
[3] The University of Tokyo

1 Introduction

This study examines a co-creative decision-making method of artifactual systems for creating effective solutions under incomplete conditions by introducing bounded rationality as a characteristic of agents.

Artifactual systems naturally reach decisions rationally or optimally to attain their purposes, as seen in a typical example, a manufacturing system. Recently however, situations surrounding artifactual systems such as manufacturing systems are becoming increasingly complex through various factors such as diversification of consumer preferences and economic fluctuation. Limitations of top-down approaches have been pointed out under such complex conditions [1]. Therefore, systems must be designed to use a bottom-up approaches that are flexible, robust, and adaptive rather than optimal.

In the engineering field, it has been common sense to eliminate effects of human decisions when artifactual systems are designed because human decisions are regarded as being unreliable and might disturb system optimization, as expressed in the commonly used term "human error". However, systems' decision-making problems in the real world are closely enough related to human factors that their influence cannot be ignored or eliminated. System design methods that actively introduce human beings are desired. The characteristic of human decision-making of not being perfectly rational or optimal has been designated as "bounded rationality" by Simon in the field of economics [2].

This study examines a decision-making method of artifactual systems based on a bottom-up approach to produce effective solutions under incomplete conditions with explicit consideration of bounded rational agents. Section 2 proposes the "co-creative decision-making" approach based on a multi-agent approach, which is an important bottom-up approach.

Furthermore, this study asserts that agents' bounded rationality has positive aspects to its use. Actual human beings have the innate ability to produce decisions flexibly on the grounds of their complexity and bounded rationality instead of their abilities to derive optimal solutions and act with perfect rationality. Section 3 describes bounded rationality to introduce it as a positive characteristic of agents. Section 4 shows a preliminary computer simulation with the simple characteristic of bounded rationality, for confirming bounded-rational agents' effectiveness.

2 Co-Creative Decision Making

Co-creative decision-making is "collective decision-making that creates an effective solution, heretofore unattained by independently-acting agents, as a whole system that allows interaction among acting agents for synthesis of an artifact." Agents are not only autonomous machines: they are also human beings and organizations such as companies. Various intercultural or interregional co-creations are expected to arise from considering the interactions between artifact and artifact, human being and artifact, human being and human being, organization and organization, and so on. Figure 1 shows that a co-creative decision-making system comprises multiple agents who act on a local environment and have their own purposes. Agents create behavioral solutions to achieve their own purposes under a certain environment. Neither special agents, which control all other agents, nor designers, who instruct agents directly from outside the system, exist.

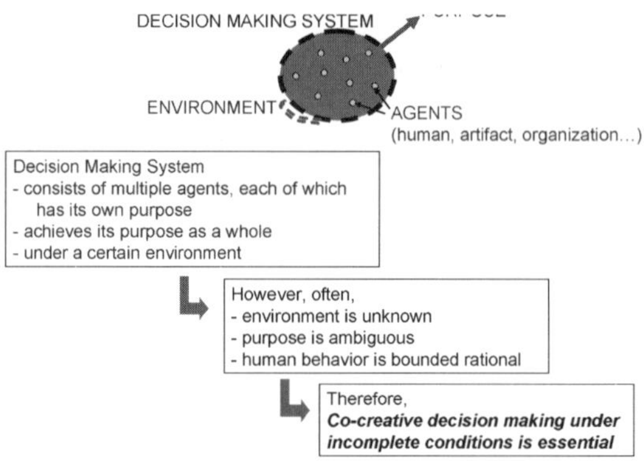

Fig. 1. Concept of co-creative decision-making

Figure 1 depicts the principal factors of incompleteness of conditions, which complicate decision-making problems in the real world: (factor-A) environmental uncertainty, (factor-B) ambiguity of purpose, and (factor-C) agents' irrationality. This study is positioned as research that specifically addresses a third factor: (factor-C) agents' irrationality.

3 Modeling Bounded-Rational Agents

3.1 Traditional Discussion of Agents' Rationality

In the field of economics, limitations or drawbacks of "perfectly rational" agent models have been indicated; perfectly rational agent models differ markedly from observed practices of human agents. For that reason, "bounded rationality" is becoming well established as an alternative. Bounded rationality is a central theme in behavioral economics. Behavioral economists list the three points as mistaken assumptions of a perfectly-rational model: (R-1) unlimited knowledge (information) and unlimited cognitive ability; (R-2) complete self-control; and (R-3) completely egoistic intentions.

Correspondingly, being bounded rational can be described as follows: (BR-1) having limited knowledge and limited cognitive ability; (BR-2) being sometimes out of control or simplistic; and (BR-3) being sometimes altruistic.

These three characteristics remain analytic descriptions. These characteristics must be correlated with the artificial agent model.

3.2 Artificial Agent Model With Bounded Rationality

Russell and Norvig reinforced the artificial agent model (Fig.2) with four requisites; "PAGE" - Percepts, Actions, Goals, and Environments [3]. An agent who exists in certain environment E with a certain goal G to achieve performs certain action A, based on information it percepts (P). Additionally, in co-creative decision-making systems, agents have inner structure (IS) because they are not mere elements, but decision-makers.

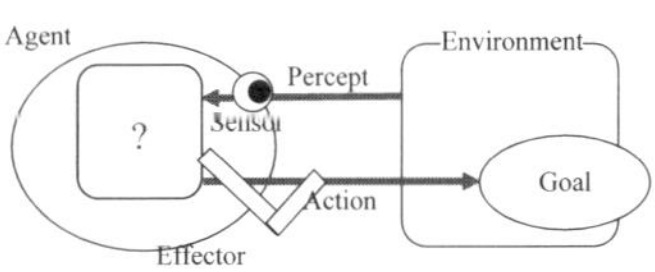

Fig. 2. Artificial agent model

Next, we revisit the factors of incompleteness of conditions mentioned in Section 2 – factors A, B, and C – which complicate decision-making problems in the real world. Corresponding with the PAGE model, factor A represents incompleteness of information about E, or inclusion of some fluctuation in E. Similarly, factor B can show incompleteness of information about G, or inclusion of some fluctuation in G. Both E and G are defined outside the agent. The remaining requisites – P, A, and IS – are defined inside the agent or are related to the inside of the agent. Because it is clear that factor C is the description linked to the inside of agents, this factor can be interpreted as incompleteness of information about P, A, or IS. Regarding BR-1, even if sufficient information to make a decision were available to an agent, it might be impossible to obtain and retain all information because of the agent's limited cognitive ability or memory. This characteristic can be modeled as a limitation or fluctuation of P. Moreover, even if an agent were able to make decisions perfectly-rationally using sufficient information, physical constraints might prevent it from executing the directive as planned. This characteristic can be modeled as a limitation or fluctuation of A. Therefore, part of an agents' bounded rationality is expressed as limitation, fluctuation, or incompleteness of P and A. Other bounded rationality behaviors – e.g., not using all available information, illogical thought, or altruistic behavior – are expressed as fluctuations of IS.

Numerous real-world decision-making problems are resolved under situations in which information about environments or goals are incomplete, and perceptions or actions are constrained. Correspondingly, this study emphasizes that actual human beings can produce decisions flexibly and adaptively. Fluctuations in IS introduce bounded rationality into artificial agents and thereby contribute to improving the system's performance.

4 Computer Simulation

4.1 Simulation Overview

We constructed a computer simulation to verify that the system's performance might be improved by introducing bounded-rational agents. As the first step, bounded-rational agents were modeled as sometimes acting randomly, which is the simplest characteristic in bounded rationality.

The simulation was modeled based on the Ant System (AS), which was inspired by observations of actual ant colonies and their inhabitants' complexly structured behavior. In AS, ants are agents with extremely simple capabilities. Consequently, to some degree, they mimic real ants' behavior.

The simulation subsumes a discrete two-dimensional space as the environment and a discrete-time model. Some food sources exist at unknown positions. The swarm of ants is a system, whose purpose is to gather all food into the nest rapidly. Each ant seeks to gather as much food as possible.

Initially, all ants are at the nest. Each apparently behaves randomly at first. At some time, one or some agents will find food. The ant that finds carries it to the nest while releasing a pheromone. In the process, the path by which the ant passed becomes a pheromone trail; the pheromone diffuses because it is volatile. A rational ant that detects the pheromone is attracted because it will reason that the pheromone trail leads to food. However, even if a bounded-rational ant detects the pheromone, it does not always follow the trail; it might continue to behave randomly with probability P. A bounded-rational ant might thereby miss opportunities to find food, but it might instead find opportunities to develop new food sources. Rational ants will gather around a discovered food source. However, food provided by each source is eventually exhausted. Ants must continuously seek new food sources. In such a case, introducing several bounded-rational ants will enhance the system performance. In this simulation, the ant behavior model and the pheromone formulation are identical to those described in [4].

4.2 Simulation 1

The first simulation was executed by: (a) changing the ratio of bounded-rational ants with $P = 1.0$ – bounded-rational ants behave randomly whenever they detect a pheromone; (b) changing P when all ants behave according to a bounded-rational mode with the same P.

We counted execution time steps from the start time to the time when all food sources were exhausted. In both simulations (a) and (b), the execution time steps were shortened: the system performance was improved by introducing bounded rationality. In this paper, the graphs of the results are omitted because of limited space. As important knowledge yielded by the results, the system performance improved dramatically, especially when all agents are bounded-rational with $P = 0.3, 0.5$. We infer that moderate randomness allows the swarm to avoid entrapment in local solutions.

4.3 Simulation 2

In this behavior model, perfectly-rational ants linger or go to the position with the highest pheromone density in the visual field. Ants with wide visual fields cannot leave a position with a known food source, even after a food source is gone; they must wait until the pheromone evaporates completely. Instead of seeking food far away, they will often linger at the position of an exhausted food source.

The second simulation confirmed the effect of the ants' visual field. Parameters used in the simulation are listed below:

Size of environment, 100*100; Number of food sources, 30; Amount of food in each food source, 20; Number of agents, 20; Visual field, variable; Random probability P, variable.

Fig. 3 shows the result. This result qualitatively shows that size of the visual field correlates with the randomness of the system: (a) wide vision improves the system performance when P is high; and (b) wide vision worsens the system performance when P is low. The visual field determines the amount of information each agent can use for its decision-making. However, information from the pheromone field is past information. Ants with a wide visual field cling to past information. Therefore, especially under a changing situation, their decision might be invalid. On the other hand, random probability P indicates the probability of acting by using information the agent has on its own will.

This result implies that an appropriate amount of information for each agent's decision-making might improve the system's performance.

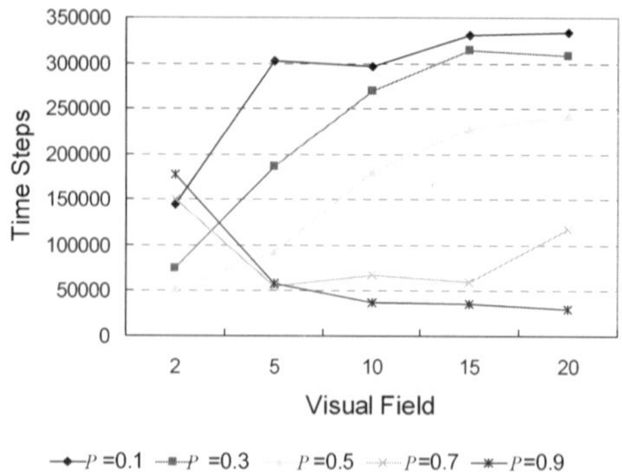

Fig. 3. Transition of average execution time steps according to random probability P and the ants' visual field

5 Conclusion

The study presented a co-creative decision-making method for creating effective solutions in artifactual systems. Herein, starting with the concept of co-creation decision-making, three principal factors of situational incompleteness were described. This study particularly addressed positive aspects of agents' bounded rationality. Bounded rationality was described by corresponding with the artificial agent model, through discussion of definitions from the field of economics. The preliminary computer simulation was based on the Ant system for verifying the validity of introducing bounded-rational agents. Random action selection was introduced as the first simple example of bounded rationality. These results demonstrate the effectiveness of introducing moderate bounded rationality into artifactual systems.

References

1. Ueda K, Markus A, Monostori L, Kals HJJ, Arai T (2001) Emergent Synthesis Methodologies for Manufacturing. Annals of CIRP 50/2:535–551
2. Simon HA (1982) Models of Bounded Rationality. 2, Cambridge, MA, MIT Press
3. Russell SJ, Norvig P (1995) Artificial Intelligence: A Modern Approach. Prentice Hall
4. Nakamura M, Kurumatani K (1999) Mechanism for Changing the Foraging Behavior in an Ant Colony Model. Complexity International 6

A Study on Virtual Market for Pareto Optimal Mediation in Economic Society

Toshiya Kaihara[1] and Susumu Fujii[2]

[1] Kobe University
[2] Kobe University

1 Introduction

Market-oriented programming is a new approach to design and implementation of resource allocation mechanisms in computer systems [1]. It has its roots in different disciplines, such as economics and computer science, in particular the area of multi-agent systems. There are our previous researches to apply market-oriented programming into resource allocation problems [2, 3]. And the Pareto optimality in market-oriented programming was left into microeconomics, and never tried to be proved in these approaches using multi-agent programming. In this paper we construct a Walrasian type virtual market, that is a principal market model in microeconomics, and try to confirm the Pareto optimality in our market model by comparing the solutions with a conventional analytic approach, named fixed-point algorithm.

2 Walrasian Virtual Market

There exists a market-oriented programming to construct a computational market (i.e. virtual market), which consists of several heterogeneous agents [1]. Definitions of the virtual market are based on general equilibrium concept in perfect competitive market, and that means it satisfies a necessary condition of Walrasian type virtual market.

Supply / demand functions represent agent's willingness to sell / buy resources, respectively. They are defined as the relationship between price and quantity of the trading resource. Let $Pt(s)$ be the price of resource s at time t. $\alpha_t ms$ and $\beta_t ns$ represent the supply function of supplier m on resource s at time t and the demand function of demander n on resource s at time t, respectively. The bidding mechanism computes an equilibrium price in each separate market. It involves an iterative adjustment of prices based on reactions of agents in the market. Agent s submits supply and demand functions ($\alpha_t ms$ and $\beta_t ns$) and the auction adjusts individual prices to clear, rather than adjusting the entire price vector by some increment. The mechanism associates an auction with each distinct resource. Agents act in the market by submitting bids to auctions. In this paper bids specify a correspondence between prices and quantities of the resource that the agent offers to demand or supply as a basic study. Given bids from all interested agents, the auction derives a market-clearing price.

3 Economic Agent

We describe demand agent (i.e. consumer) as c_m $(m = 1, 2, \cdots, M)$, and supply agent (i.e. producer) as s_n $(n = 1, 2, \cdots, N)$. The number of kind of goods is assumed as I in our Walrasian VM.

3.1 Demand Agent

Demand Utility

Suppose demand agent c_m has utility function u^{c_m}, which is described with in Eq.(1). Here $x_i^{c_m}$ represents the demand quantity for resource i:

$$u^{c_m} = a^{c_m} \prod_{i=1}^{I} (x_i^{c_m})^{b_i^{c_m}}$$
$$\text{where} \sum_{i=1}^{I} b_i^{c_m} = 1 \quad (0 < a^{c_m}, b_i^{c_m}) \tag{1}$$

In this paper we adopt Cobb-Douglas function [4] as a demand function described in Eq.(1), because the Cobb-Douglas function is one of the primitive functions in microeconomics, which handles economical scale in the market by index constant b.

Budget

Budget of demand agent c_m is formulated by initial quantity of resource (i): $e_i^{c_m}$, and their price: p_i as follows:

$$B^{c_m} = \sum_{i=1}^{I} p_i e_i^{c_m} + r^{c_m} \tag{2}$$

In this equation r^{c_m} represents supplier's profit, which suppliers return to demanders under zero-profit conditions in the general equilibrium theory [5].

Bidding Functions

Demand agents send their bid to their target resources in the market, and the bid is formulated as demand function. The function is obtained as the optimal solution as maximising problem of Eq.(1) under the constraints described in Eq.(2). The following demand function is calculated by Lagrange's method of (indeterminate) multiplier in this research.

$$x_i^{c_m}(p_i) = \frac{b_i^{c_m} B^{c_m}}{p_i} \quad i = 1, 2, \ldots, I \tag{3}$$

We assume demander c_m supplies all the initial resources into the market according to the principle of microeconomics, and demander's supply function of resource j is defined as follows:

$$y_j^{c_m} = e_j^{c_m} \tag{4}$$

3.2 Supply Agent

Production Function

As described in demand agent definitions, Cobb-Douglas function is basic functions which handles economical scale in the market easily. In microeconomics production function is assumed to be concave function, and that means market prices are established at a predictable level in the general equilibrium theory in concave shape production function.

We also formulate production function of supply agent sn to resource j as Cobb-Douglas function to satisfy the assumption, shown as Eq.(5). Cobb-Douglas function is defined as a concave function in $0 < \beta < 1$ in this equation.

$$y_j^{s_n} = \alpha^{s_n}(x_i^{s_n})^{\beta^{s_n}} \quad (0 < \alpha^{s_n}, 0 < \beta^{s_n} < 1) \tag{5}$$

Profit

According to microeconomics assumption, supply agents have no initial resources. They can earn their profit π^{s_n} by producing value added resources from purchased resources. The profit function is defined as follows:

$$\pi^{s_n} = p_j y_j^{s_n} - p_i x_i^{s_n} \tag{6}$$

Bidding Functions

Supply agents send supply functions to production resources, and demand functions to purchase resources, respectively. They maximise their profit by solving maximising problem of Eq.(6) under the constraint in Eq.(5). We also solve the problem by Lagrange ' s method of multiplier in this research, and obtain the following demand function and supply function in Eq.(7) and (8), respectively:

$$x_i^{s_n}(p_i) = \left(\frac{p_i}{\alpha^{s_n}\beta^{s_n}p_j}\right)^{\frac{1}{\beta^{s_n}-1}} \tag{7}$$

$$y_j^{s_n}(p_j) = \left(\frac{(p_i)^{\beta^{s_n}}}{\alpha^{s_n}(\beta^{s_n})^{\beta^{s_n}}(p_j)^{\beta^{s_n}}}\right)^{\frac{1}{\beta^{s_n}-1}} \tag{8}$$

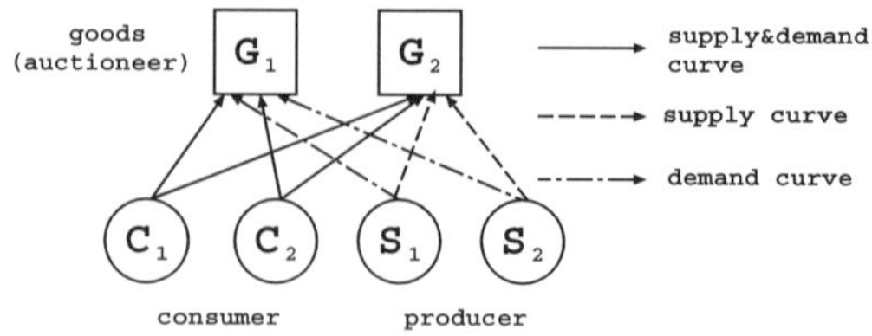

Fig. 1. Experimental market

Table 1. Initial conditions of consumers

Agent	Utility function(u^{c_m})	Endowment ($e_1^{c_m}, e_2^{c_m}$)	Utility value
c_1	$1.0(x_1^{c_1})^{0.8}(x_2^{c_1})^{0.2}$	(30.0,80.0)	36.50
c_2	$1.0(x_1^{c_2})^{0.3}(x_2^{c_2})^{0.7}$	(70.0,20.0)	29.12

Table 2. Initial conditions of producers

Agent	Input goods	Output goods	Production function
s_1	goods 1	goods 2	$y_2^{s_1} = 2.0(x_1^{s_1})^{0.7}$
s_2	goods 1	goods 2	$y_2^{s_2} = 3.0(x_1^{s_2})^{0.5}$

4 Experimental Results

Experimental 2-producer-2-consumer market is illustrated in Fig.1. In this market 2 kinds of goods, goods 1 and goods 2, are also traded. Experimental values of each parameter (i.e. initial value set) in the model are shown in Table 1(consumer agent) and Table 2 (producer agent). Consumer agents send their bids as supply and demand functions to both the goods homogeneously. Producer agents send their demand and supply function to goods 1 and goods 2, respectively. That means they both produce goods 2 from goods 1 in this market. Simulation results on price changes of each goods in the Walrasian VM are shown in Fig.2. In this model we obtained equilibrium price vector $\hat{\boldsymbol{p}} = (\hat{p_1}, \hat{p_2})$ as follows:

Equilibrium price vector: $\hat{\boldsymbol{p}} = (0.95673, 0.98543)$
Normalised equilibrium price vector: $\hat{\boldsymbol{p}} = (0.49261, 0.50789)$

And final resource allocation in consumer agent and producer agent are shown in table 3 and 4, respectively.

It has been observed that the total consumer's utility is increased in table 3 compared with table 2, because the producer's profit is returned to consumer agents in this market under the zero-profit conditions in the general equilibrium theory. The utility of agent c_2 is increased by 97.3 % especially. That is because of the agent's preference for goods 2 as well as the zero-profit conditions. In this market all the producer agents supply goods 2, and that

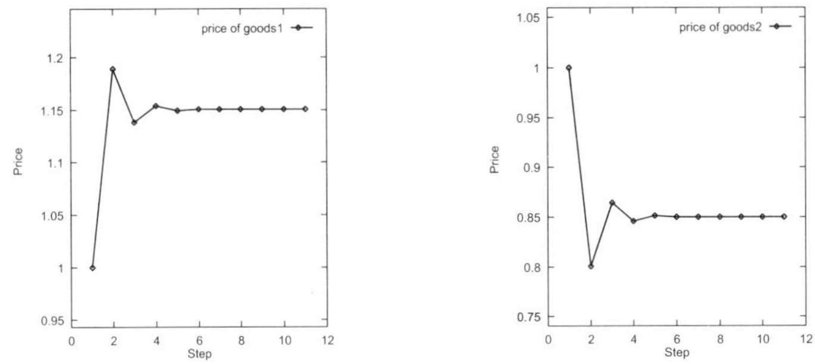

Fig. 2. Price transition in trade market

increases the utility of agent c_2. The utility of agent c_1 is increased by 57.4 %, less than c_2, because of its stronger preference for goods 1.

Table 3. Equilibrium conditions of consumers in trade market

Agent	Consumption $(x_1^{c_m}, x_2^{c_m})$	Utility value
c_1	(71.96,24.35)	57.94
c_2	(25.69,81.14)	57.46

Table 4. Equilibrium conditions of producers

Agent	Production	Input	Profit
s_1	2.16	1.12	0.55
s_2	3.32	1.23	1.41
Total	5.48	2.35	1.96

We applied fixed point algorithm in this market model to confirm Pareto optimality of the VM solutions. In using Scarf's algorithm to find such a fixed point, the unit simplex is divided into a finite number of smaller simplices (i.e. grid size) [6]. In this paper the grid size is set to 100,000 for the precise comparison. We obtained equilibrium price vector $\hat{\boldsymbol{p}} = (\hat{p_1}, \hat{p_2})$ by the fixed point algorithm as follows:

$$\text{Equilibrium price vector:} \quad \hat{\boldsymbol{p}} = (0.49262, 0.50789)$$

It is obvious that the equilibrium price set obtained by Walrasian VM is almost equivalent to the one from the fixed point algorithm, and that means VM solutions have been confirmed to be converged into Pareto optimal. The small difference is caused by the grid size of the fixed point algorithm.

We compared the calculation time between VM approach and fixed point algorithm. CPU consumption time (second) for each approach in this market model is as table 5.

Table 5. CPU consumption time

Method	CPU time (second)
VM	0.0017
Fixed point algorithm	12.025

VM approach is obviously more than 7,000 times as fast as the analytic approach in this model. It has also been proved that the proposed VM based approach is much more practical in terms of calculation time to obtain Pareto optimal solutions in resource allocation problems.

By the computer simulation, we have confirmed that Walrasian VM takes advantage of the market analogy into resource allocation problem, and that leads to effective search of Pareto optimal solution for supply chain management.

5 Conclusions

Market-oriented programming is a new approach to design and implementation of resource allocation mechanisms in computer systems, and we newly proposed a Walrasian Virtual Market (VM) approach with microeconomics based market-oriented programming. Firstly we mentioned our general concept to apply VM into resource allocation problems, and explained general idea of Walrasian market model in economics. Then central aspects of market-oriented programming are investigated and some new insights are presented. After a brief explanation of the analytical approaches, named fixed-point algorithm, we defined agent behaviour based on Walrasian market-oriented programming. As a basic study, we analysed its Pareto optimality by computer simulation experiments, and it has been confirmed that our approach is efficient both in Pareto optimality and calculation performance.

References

1. Wellman MP (1996) A Market-Oriented Programming Environment and its Application to Distributed Multi-commodity Flow Problems, ICMAS-96, 385–392
2. Kihara T (1999) Supply chain management based on market Mechanism in virtual enterprise, Infrastructures for Virtual Enterprises, Kluwer Academic Publishers, Boston, 399–408
3. Kihara T (2001) Supply chain management with market economics, International Journal of Production Economics, Elsevier Science, Vol. 73, Issue 1, 5–14
4. Kreps DM (1990) A Course in Microeconomic Theory, Harvester Wheatsheaf, New York
5. Shoven JB, Whalley J (1992) Applying General Equilibrium, Cambridge University Press, New York
6. Scarf HE (1973) The computing of Economics Equilibria, New Heaven, Yale University Press

Empirical Tests of Exchange Rate Theory

Eric O'N. Fisher *

California Polytechnic State University

1 Introduction

There have been very few direct applications of experimental techniques in macroeconomics for two main reasons. First, macroeconomics is about the interaction between markets, and it is not easy to design an elegant treatment that gets at the essence of how a national economy functions. Second, there is a vestigial prejudice that favors econometrics over putatively unorthodox empirical approaches. Still, the Nobel Prize-winning economist Edward Prescott is alleged to have said, "Don't regress; progress!"

This paper takes that methodological exhortation seriously, and it pays homage two recent Nobel Prizes in economics. Following the pioneering work of Vernon Smith and others, I use the laboratory to study some foundations of international finance. Following in the footsteps of Fynn Kydland, Edward Prescott, and others, I eschew esoteric econometric techniques and design direct empirical tests of the three most basic elements of exchange rate theory: purchasing power parity, covered interest parity, and uncovered interest parity.

My research establishes two points. First, each of these three elements of exchange rate theory fares well in the laboratory. Second, not-traded goods and non-stationary domestic prices do cause deviations from simple theoretical predictions, but a non-stationary environment has a more significant effect than does not-traded goods in explaining why simple predictions of exchange rate theories break down.

Perhaps the most celebrated article in the literature on exchange rate econometrics is Meese and Rogoff [9]. Those authors set the standard for a vast literature in empirical international finance. The entire literature on

* The author thanks the National Science Foundation for grants SES-9870874 and SES-0111315 that made this work possible.

experimental economics and exchange rates consist of four papers: Arifovic [1], Noussair, Plott, and Riezman [10], Fisher and Kelly [4], and Fisher [3].[1]

How does my work contribute to the larger literature on empirical models of exchange rates? My simple treatments may not capture the full complexity of the international monetary system, but experimental economics does complement traditional research using applied econometrics. My results reinforce the notion that both not-traded goods and non-stationary prices contribute to the empirical failure of exchange rate models, but this experiment shows clearly that non-stationary prices have a stronger effect.[2]

2 Elements of Exchange Rate Theory

There are three elements that are at the foundation of any theory of international finance. The first is purchasing power parity; nominal exchange rates adjust, at least in the long-run, so that comparable baskets of goods from different countries will cost the same when denominated in a common currency. The second is covered interest parity; most foreign exchange trading among the major money center banks involves simultaneous purchases of spot foreign exchange and resale on the forward market. Since this is a synthetic transaction that entails no risk, it defines the foreign exchange forward premium in terms of home and foreign risk-free rates of interest. The third element is uncovered interest parity. This element states that risk-neutral traders would be willing to hold an open position in foreign exchange if expected dpreciation is less that the difference between home and foreign interest rates.[3]

3 The Experimental Design

Table 1 summarizes the experiment's design and also gives goodness-of-fit statistics. Its columns are the three different treatments, and its rows capture different elements of exchange rate theory. Since the experimental design constitutes a two-way layout with five blocks and three treatments, it is easy to examine treatment effects. [4] A general test has the interpretation that either not-traded goods or non-stationary prices cause different outcomes for at least one of the three elements.

[1] Let me put this number in perspective. A search of EconLit in March 2005 found 19132 hits for the keywords "exchange rates," 1009 hits for "purchasing power parity," and 1182 hits for "money demand."

[2] Using an intuitive statistical decomposition, Engel (1999) showed that not-traded goods prices accounted for almost none of the variability of the U.S. bilateral real exchange rate against several major countries over a long period.

[3] The motivated reader can consult Froot and Rogoff [5] or Frankel and Rose [6] for a lengthier discussion of these ideas.

[4] See Hollander and Wolfe [8], chapter 7.

		Macroeconomic Environment		
		Control	Not-Traded Goods	Non-Stationary Prices
Element of Exchange Rate Theory	Purchasing Power Parity	0.13	0.28	0.33
		0.08	0.08	0.26
	Covered Interest Parity	0.02	0.07	0.26
	Uncovered Interest Parity	0.15	0.38	0.19
		0.05	0.22	0.67

Table 1. Goodness of Fit Statistics, Mean Absolute Percentage Errors

4 Experimental Procedures

The sessions were held in between October 2000 and June 2002 at The Ohio State University, and they were conducted by hand, not by computer.

The experimenter auctioned off one unit of currency to each of the four highest bidders; hence foreign exchange was in fixed supply. The experimenter sold domestic goods and foreign goods at fixed prices; hence the markets for commodities exhibited perfectly elastic supply. Domestic currency and foreign currency were not worth anything in and of themselves, but one could buy domestic goods only with domestic currency, and foreign goods only with foreign currency. Thus the experiment imposes two cash-in-advance constraints. In order to buy foreign commodities, a subject had to purchase foreign exchange by bidding in units of domestic currency. The foreign exchange markets were third-price auctions, and thus it is a Nash equilibrium for each subject to bid the common value for a unit of foreign currency.

The foreign exchange market cleared twelve times in the sessions in the first and second blocks of Table 1, and twenty-four times in all the other sessions. The subjects earned about $25 on average. The currencies were different kinds of play money. The instructions made the two cash-in-advance constraints very explicit, and they also go into detail about how the call market works. After having read the instructions together with the experimenter aloud, all the subjects answered a series of questions to make sure that they understood the incentives inherent in each design. [5]

5 Market Equilibrium and Experimental Results

The no-surplus Nash equilibrium forms the theoretical predictions for the model in every session. There are deviations from the theoretical predictions,

[5] The instructions can be found at http://economics.sbs.ohio-state.edu/efisher/ppp.

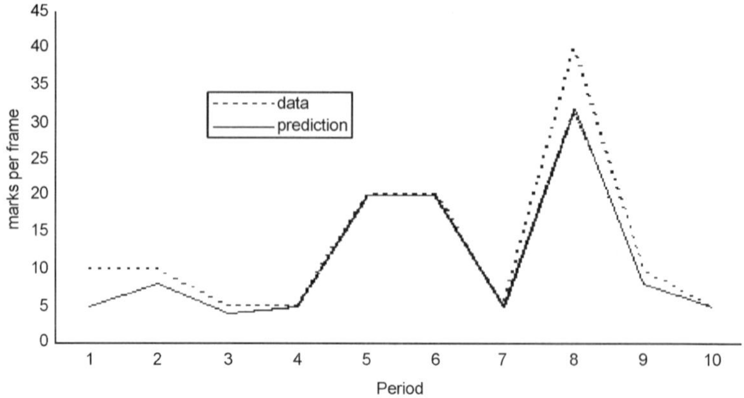

Fig. 1. Purchasing Power Parity, 5 October 00

but the five sessions that constitute the control treatment have remarkably accurate theoretical predictions. The sessions from the treatments with not-traded goods and with non-stationary prices do show greater deviations from the theoretical predictions than those in the control group.

The results are best summarized by graphs. Figures 1 through 3 show the predictions and the actual data for the purchasing power parity five. Figure 1 is the control treatment, Figure 2 is the treatment with not-traded goods, and Figures 3 is the treatment with non-stationary domestic prices.

Three general observations about these fifteen sessions are salient. First, the purchasing power parity, covered interest parity, and uncovered interest parity fare very well. Second, the control sessions exhibit remarkably accurate theoretical predictions. Third, both not-traded goods and non-stationary domestic prices cause deviations from the theoretical predictions, but the biggest deviations from the theoretical predictions seem to occur because of the non-stationary environment.

Table 2 gives the mean absolute percentage errors of the data from the model's predictions. The natural non-parametric test for treatment effects is Friedman's [7] rank sum test. One can reject the null hypothesis of no treatment effect for a test of size 1%. [6] The statistics in the third column of Table 2 are much larger than those in the second column. That table presents

[6] In both cases, the Friedman statistic is 8.4, and it has a p-value of .008. There were no ties; the data shown in Table 4 have been rounded for ease of exposition. The exact distribution of this statistic is given in Hollander and Wolfe (1973), Table A.15. It has an asymptotic chi-squared distribution whose degrees of freedom are one less than the number of treatments.

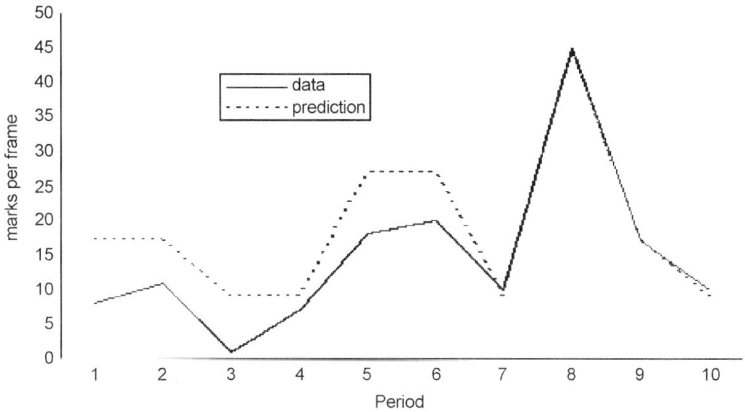

Fig. 2. Not Tradeed Goods Purchasing Power Parity, 8 February 2002

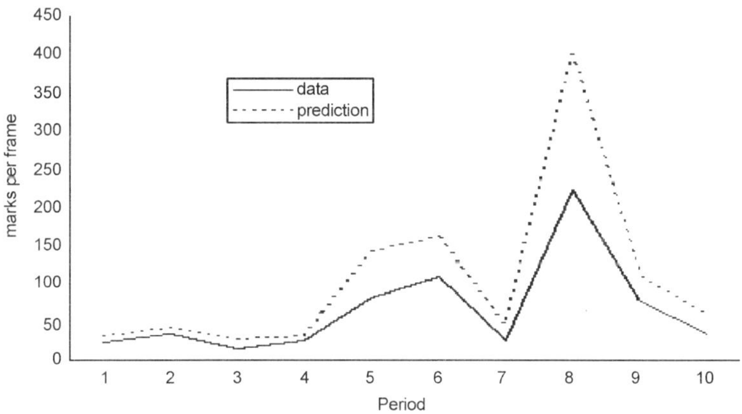

Fig. 3. Not - Stationary Purchasing Power Parity, 1 February 2002

prima facie evidence that the largest departures from the theory have to do with non-stationary domestic prices.

6 Conclusion

The experiment shows that all three elements of exchnage rate theory hold in a simple environment. Exchange rate theory breakd down primarily because of trending prices. The find that not th treatment with traded goods has a much weaker effect than the one with trending prices is the main contributions of this experiment.

I would like to conclude with an exhortation for more experimental research in macroeconomics and international economics. The hallmark of a science is the replication of empirical results. I have shown that the elements of exchange rate theories are vindicated in the laboratory. But I have also used a controlled environment to indicate perhaps that the broad empirical failure of many exchange rate models may have to do with the disparate secular inflation rates that the major industrial have experienced.

References

1. Arifovic, J (1996) The behavior of the exchange rate in the genetic algorithm and experimental economies, Journal of Political Economy 104: 510-541.
2. Engel C (1999) Accounting for U.S. real exchange rate changes, Journal of Political Economy 107: 507-538.
3. Fisher EO'N (2001) Purchasing power parity and interest parity in the laboratory, Australian Economic Papers 40: 586-602.
4. Fisher EO'N and Kelly FS (2000) Experimental foreign exchange markets," Pacific Economic Review 5: 368-388.
5. Frankel JA, Froot KA (1995) Perspectives on PPP and long-run real exchange rates, Chapter 32 in Handbook of International Economics, Volume 3, (Grossman GM and Rogoff K, eds.), Elsevier, Amsterdam.
6. Frankel JA, Rose AK (1995) Empirical research on nominal exchange rates," Chapter 33 in Handbook of International Economics, Volume 3, (Grossman GM and Rogoff K, eds.), Elsevier, Amsterdam.
7. Friedman M (1937) The use of ranks to avoid the assumption of normality implicit in the analysis of variance, Journal of the American Statistical Association 32: 675-701.
8. Hollander M, Wolfe DA (1973) Nonparametric statistical methods. John Wiley: New York.
9. Meese RA and Rogoff K (1983) Empirical exchange rate models of the seventies: Do they fit out of sample? Journal of International Economics 14: 3-24.
10. Noussair C, Plott C, Riezman R (1997) The principle of exchange rate determination in an international finance experiment, Journal of Political Economy 105: 822-861.

Charting the Market: Fundamental and Chartist Strategies in a Participatory Stock Market Experiment *

László Gulyás[1] and Balázs Adamcsek[2]

[1] AITIA International Inc
[2] Loránd Eötvös University

1 Introduction

Theorists and market traders have strikingly different views about financial markets. [5] Standard theory assumes identical investors who share their rational expectations about an asset's future price. Consequently, speculation cannot be profitable, except by luck; trading volume stays low, and market bubbles and crashes reflect rational changes in the asset's valuation. In contrast, traders do speculate in practice. Also, market deviations exist and are often ascribed to market psychology. There is also an interpretation of these differences at the level of practical trading rules. If speculation works, technical rules that are based on only price or trade volume information may be useful. According to rational expectations theory, however, only fundamental strategies that relate price to fundamental value by using dividend information will yield success.

One way to study questions related to the debate above is to construct artificial markets with computational agents, using agent-based modeling (ABM). One of the first and most prominent of such models is the Santa Fe Institute's Artificial Stock Market model (SFI ASM). [4][5] In [2] and [3] we describe a participatory extension of that model, in which human traders replace some of the agents. Our experiments show that even a few human agents may significantly alter aggregate market performance. Furthermore, our results suggest that technical trading lends itself easily to inexperienced traders, but fundamental strategies perform better in the artificial stock market. There we conjectured that the latter may be due to the dominance of computational agents, who might have developed mostly fundamental strategies. In this paper we report on further experiments designed to investigate this hypothesis. We explore how extreme market deviations effect the strategies adopted by

* This research was supported by the GVOP-3.2.2-2004.07-005/3.0 (ELTE Informatics Cooperative Research and Education Center) grant of the Hungarian Government.

inexperienced traders and show that market deviations may lead participants to use technical strategies. Furthermore, we study what effect these adopted strategies, in turn, have on aggregate market behavior. We show that the chartist rules adopted by the human traders may, in fact, contribute to a rational expectations equilibrium.

2 The Participatory Santa Fe Artificial Stock Market

The SFI ASM is a minimalist model with a risk-free financial asset (e.g., Treasury bills) available in infinite supply that pays a constant risk-free return rate per period, and with a risky stock, the fundamental share value of which is unknown to the traders. Traders are computational agents that trade on a spot market, where the price per share of the stock depends on the overall buying and selling behavior of the agents. The stock, however, may also pay a dividend in money, which is predetermined, but unknown to the traders. [5]

The trading agents of the SFI ASM model try to maximize their wealth by regularly changing their portfolio. Their behavior is based on evolving rules that specify what to do when certain market conditions are met. The possible actions are buying, selling, or holding, which are dependent on a condition string representing the current state of the market. [4] [1] The symbols in the string represent market indicators. The rules may require an indicator to be true or false, or they may be indifferent in regard to some of the indicators.

The first 6 market indicators (out of the 12, as shown on Table 1) represent information used in fundamentalist strategies. In contrast, indicators 7-10 are technical (chartist) tools. The last two are zero information indicators, providing a way to check whether the agents' behavior is actually dependent on market processes.

Table 1. Market Indicators

Indicator	Market Indicator	Indicator	Market Indicator
1	$p^t rd^t > 1/4$	7	$p^t >$ 5-period moving average price
2	$p^t rd^t > 1/2$	8	$p^t >$ 10-period moving average price
3	$p^t rd^t > 3/4$	9	$p^t >$ 100-period moving average price
4	$p^t rd^t > 7/8$	10	$p^t >$ 500-period moving average price
5	$p^t rd^t > 1$	11	Always true
6	$p^t rd^t > 9/8$	12	Always false

In [2] and [3], we found that the presence of human traders yielded higher market deviations, measured as the cumulative difference between stock price and fundamental value. Also, in their post-interviews, human players reported to start with technical (chartist) strategies, then, gradually, a few of them discovered fundamental strategies.

3 Humans Modulating the Market: Experiments With Chartist and Fundamentalist Agents

In this paper, we introduce two new types of artificial traders: chartists and fundamentalists. Chartist agents evolve rules that depend on technical market indicators only. In contrast, fundamentalist agents evolve rules that only take fundamentalist indicators into account.

In the briefings preceding the experiments the market indicators received a special emphasis. Since technical indicators are generally easier to grasp, the focus of the discussion was on the fundamental ones. Then the participants were subjected to five different market environments. First, they played an introductory session among themselves without any artificial agents. Second, the subjects were confronted with an equal number of artificial chartist agents. Third, human participants consisted the 5% of the market, the remaining 95% were chartist agents. Fourth, human participants amounted for the 5% of the market as in the previous case. However, the remaining 95% was split between chartist agents (50%) and the original learning agents (45%) of the SFI ASM model. Finally, human subjects were confronted with an equal number of SFI ASM agents. The concept was to confront the subjects with markets of very high volatility (significant bubbles and crashes) and study how people who received a special briefing on fundamentalist indicators will react.

An analysis of the post-experiment questionnaires shows that our subjects, in response to being subjected to 'bubble-and-crash' markets, gradually adopted chartist strategies. In contrast, the general market behavior displayed on Figure 1 shows a tendency towards a rational expectations equilibrium during Setting 3 and 4. In the first two experiments, human participants represented a significant portion of the market (100% and 50%) and thus their trading decisions determined the aggregate behavior of the system. In particular, as the players bought up most of the stocks, the market's liquidity fell drastically. Scenario 5 shows a similar behavior.

However, in Setting 4, the market shows technical trading apparently modulated by the market. Whether this was caused by the presence of human traders or that of the original SFI ASM agents needs further consideration. Therefore, Figure 1 also provides a more detailed look at the effect of human presence in terms of cumulated market deviation (D^t, shown in millions).

The comparison of participatory runs with baseline experiments, where original SFI ASM learning agents replaced human players, for Setting 3 and 4 confirm the role of human agents in modulating the market. It is also informative that in Setting 3, starting from the second bubble, the bursts appear earlier in participatory runs. This is probably due to human subjects cashing in on the increased price level. Not surprisingly, the absolute winner of this scenario was one of the human players.

To further confirm this finding, another series of computational experiments were carried out with 95% of chartist agents. The remaining 5% was varied to use chartist (C), fundamentalist (F), or the original SFI ASM learn-

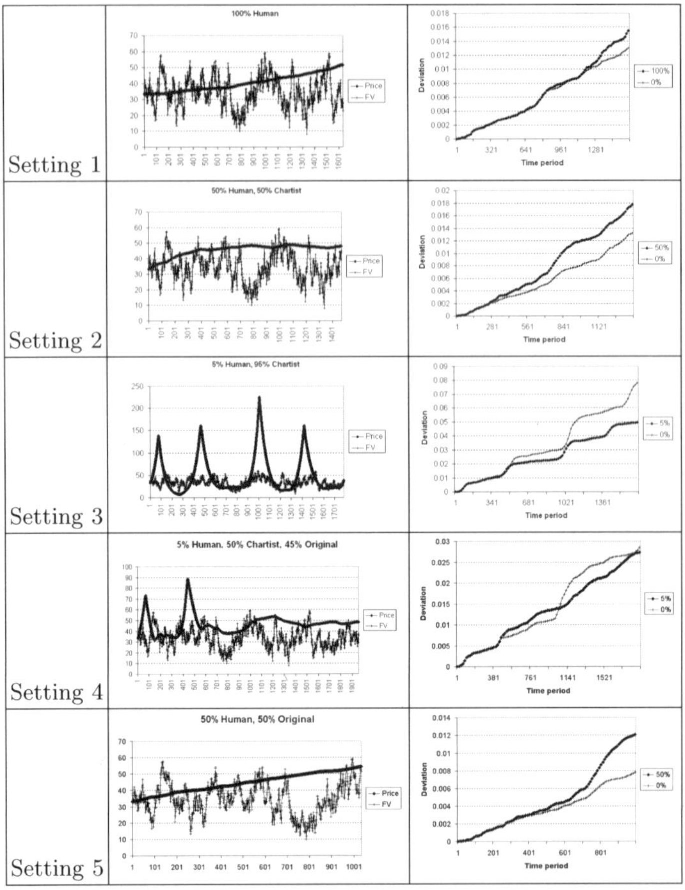

Fig. 1. Market behavior in the five experimental settings. The charts on the left show price and fundamental value vs. time. The figures on the right compare the effect of human presence to the effect of learning agents. They show cumulative deviations (D^t - shown in millions) vs. time. The two data series on the latter charts stand for participatory experiments (dotted series) and for runs where human players were substituted by SFI ASM agents (0% human participation), respectively.

ing strategy (O). We then compared the resulting market behaviors to the effect of 5% human traders. The top panel of Figure 2 summarizes our findings. The additional 5% C agents have the same effect as that of 5% O agents, since the key to the SFI ASM strategy is learning. Surprisingly, however, a 5% of F agents results in slightly increased deviation. Nonetheless, it is clear that the 5% of human-controlled agents significantly decrease market deviation in comparison to any of the previous cases. The figure also shows that human participants terminate bubbles 'early', starting from the second occurrence.

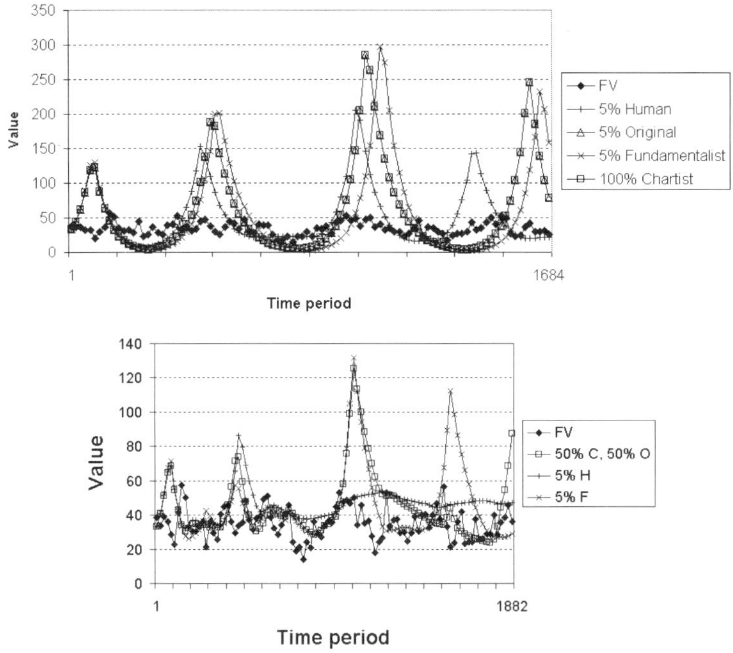

Fig. 2. Comparison of different strategies and the performance of human subjects. The charts show fundamental value (FV) and price vs. time. On the top panel, 95% of the agents applies technical rules, while on the bottom one, 50% of them are technical traders and 45% consists of SFI ASM agents.

We also compared the effects of 5% fundamentalist (F), original SFI ASM (O) and human (H) agents, in a setting where the rest of the population consisted of 50% chartist and 45% SFI ASM agents. The bottom panel of Figure 2 summarizes our findings. An equal split between C and O traders results in large market deviations, albeit in smaller ones than in the 95% C scenario. The replacement of 5% O agents with F's slightly increases the deviance. This surprising finding is consistent with those of the previous scenario. In contrast, the 5% human players, whose majority by this time followed technical trading rules, clearly decreased market deviation.

These results suggest that it is indeed human presence that moderates market deviations. Moreover, fundamental strategies, when applied in the given small percentages, seem to have a counter-intuitive effect in amplifying deviations caused by technical trading.

4 Discussion and Future Work

This paper continued our line of research exploring the effect of human traders in the early SFI ASM model. Our experiments showed human traders adopting technical strategies in response to highly volatile markets with drastic price bubbles. The effect of this experience also carried through to more stable markets. On the other hand, technical trading performed by humans had a moderating effect on aggregate market behavior in comparison to chartist-dominated markets with artificial traders only. This is a surprising finding, whose likely cause is in the specifics of the SFI ASM market and the design of its agents. More specifically, our current hypothesis is that the phenomenon can be, at least in part, explained by the learning rate of the artificial agents. That is, we conjecture that human traders were able to adapt to market situations more quickly than the artificial agents with a fixed adaptation rate. The same time, human participants' impatience on capitalizing on their gains (e.g., during the raising period of a bubble) may also explain why chartist participants actually caused bubbles to burst earlier. The detailed investigation of these issues is the subject of future work, including experiments with various adaptation rates for the artificial agents.

References

1. Ehrentreich, N. (2002) The Santa Fe Artificial Stock Market Re-Examined - Suggested Corrections. WUSTL Economics Working Paper Archive, Computational Economics Series No. 0209001. (Submitted to Journal of Economic Dynamics and Control)
2. Gulyás, L., Adamcsek, B. and Kiss, Á. (2003) Experimental Economics Meets Agent-Based Finance: A Participatory Artificial Stock Market. In: Proceedings of 34th Annual Conference of International Simulation and Gaming Association (ISAGA 2003).
3. Gulyás, L., Adamcsek, B. and Kiss, Á. (2004) An Early Agent-Based Stock Market: Replication and Participation. Rendiconti Per Gli Studi Economici Quantitativi, Volume unico, pp. 47-71.
4. LeBaron, B. (2001) A builder's guide to agent-based financial markets. Quantitative Finance, 1, 2001: 254-261.
5. Palmer, R. G., Arthur, W. B., Holland, J. H., LeBaron, B., Tayler, P. (1994) Artificial Economic Life: A Simple Model of a Stock Market. Physica D 75: 264-274.

Part VI

Between the System and Individual Behaviour

Audit Credibility and the Audit Fees: A Theory and an Experimental Investigation

Tatsuhiko Kato

Meiji University

1 Introduction

The audit fees have been recently at issue in Japan. They are too low compared with other developed countries and bring Japanese audit firms to financial difficulties. Despite of concerns, most public companies in Japan give the cold shoulder, since they want to spare any cost other than operating ones. Some people fear that this may compromise the credibility of auditing in Japan. The aim of this paper is to examine how the different audit fees influence the behavior of investors and managers.

We build a simple signaling model. Signaling can work, if the cost incurred is different from one type of agent to another. We illustrate that the wider goes the difference, the more clearly reveals oneself the distinction of behaviors among the agents. Furthermore, the magnitude of signaling cost has also an influence on their behaviors in our laboratory markets. Too much signaling cost makes the distinction less clear, since it is regarded as a waste and reduces an incentive to reveal oneself.

In the model a manager and an investor interact and there are two types of managers: dishonest and opportunistic. The dishonest type is completely effort-averse and never makes an effort. The opportunistic type trade-offs the current benefit from making no effort and the loss in reputation. In other words, he is willing to make an effort, only if he is better off. Auditing is available, but imperfect and sometimes makes mistakes. If the audit purchase is supposed to be a signal for the manager's honesty, it depends largely on the audit accuracy.

Our model predicts that the higher the audit fee is, the easier is it to obtain the separating equilibrium, in which any manager who makes no effort no longer purchases an audit. The experiment is conducted based on this model. Three laboratory markets are made: whether auditing is unavailable or available but at two different audit fees. The audit report's accuracy is unchanged regardless of the audit fees, as is widely considered in Japan. Contrary to our prediction, investors invest less often at the high audit fee. Managers also make an effort less frequently. They try to cheat more often the investors, who turn out to be defensive and refrain from the investment.

Similar experimental research is done by Dejong et al. [1], Dopuch et al. [2], Kachelmeier [4], Dopuch and King [3], and Wallin [8]. The more advanced studies, whose objective is mainly the auditors reporting behavior, are done by Mayhew et al. [6] and Mayhew and Pike [7], but they have never focused on the audit fees.

2 The Model

Suppose that a manager sells an asset to an investor. The manager can make an effort or no effort. Whenever he makes an effort, he incurs the cost ($e = 5$). The investor can buy this asset or boycott it. If he would like to buy it, he should always pay p ($p = 10$) to the manager. The liquidation value of asset will amount to 15, if the quality is high, that value will be only 5, if the quality is low. Only the manager knows whether he has made an effort or not. The audit purchase can transmit information from a manager to investors and solve the information asymmetry.

We propose a signaling model. Auditing is available, but only the manager can voluntarily purchase it at the cost of C. Two parameters α and β are provided. The former denotes how often the manager's intention, whether good or bad, will be fruitful. The latter indicates how often a correct report will be provided on the audit purchase. The parameters α and β are uniformly distributed on [0,1]. We assume that an incorrect report is provided, only if the asset quality is actually low. It means no mistake made in the audit report when the asset quality is high.

First we examine the pooling equilibrium, in which the opportunistic managers behave honestly. In other words, his incentive compatibility constraint or self-selection constraint will be satisfied. All we have to do is to make his expected payoff a little bit more than when he makes no effort and still purchases an audit. Except for the completely effort-averse type, he knows that he will be better off when he makes an effort. Hence we have

$$\alpha \geq \frac{1}{4\beta} + \frac{1}{2} \tag{1}$$

In this equilibrium the dishonest type or the completely effort-averse type can still purchase an audit. We, therefore, examine the separating equilibrium, in which those who purchase an audit will certainly make an effort. All we have to do is to set the expected payoff of dishonest type as small as when he makes no effort and neither does purchase an audit, so that the investor doesn't invest.

$$\alpha \geq \frac{10 - C}{10\beta} \tag{2}$$

This condition is dependent of C and it defines the individual rationality or participation constraint for the dishonest type as well. Suppose that α and β are stable, the higher is C, the easier is it to obtain the separating equilibrium. Our model is based on one shot game. The repetition of trade could lead to another conclusion, but it can serve as a basis on discussions in the repeated game.

3 The Experimental Investigation

3.1 The Experimental Design

The examination is also made in a laboratory setting. All subjects are undergraduate students in Japan. By using computer terminals, we conduct a matching game played by a pair of subjects: a manager and an investor. However, throughout the experimentation we are using the words "seller" and "buyer" instead of "manager" and "investor" in order to be less idiosyncratic. Three laboratory markets are created. In one market auditing isn't available, while in other two each manager can voluntarily purchase it. The subjects experience all three markets.

The asset quality is always disclosed high. Each manager can make an effort to increase the probability of reaching the high quality, but he doesn't know exactly which quality he has achieved. Each investor knows immediately whether or not the manager has purchased an audit and the audit report is also published at once. An incorrect report will be provided, only if the asset quality is actually low. In all markets the investor's payoff is immediately known after the choice was made, whereas the investor can observe imperfectly in which period the manager has made an effort. All he can do is to make a guess according to his payoff. Trade is repeated in the 20 periods.

The experimental design and examination are made according to our signaling model. One of our previous studies(Kato [5]) is conducted at $\alpha = 5/6$, $\beta = 4/5$, $C = 1$, that is, the boundary of pooling regions. The parameter α and β are set up in such a way that the expected benefit is break even between the honest and dishonest behavior on the audit purchase, thereby measuring its pure effect on the behavior of managers and investors. The results are mixing, and hence we set up $\alpha = 5/6$ in all three markets and $\beta = 4/5$ in two markets, where auditing is available, but change the audit fees, either low (C =1 yen) or high(C = 1.5 yen). Only those figures are announced to the participants beforehand.

3.2 The Results

Our 4 hypotheses are:

H1: The number of effort made by managers is higher when auditing available.

Table 1. Results (Effort and Investment), **: Significantly Different at p=0.01, *: Significantly Different at p=0.05 (one tail), SD: Standard Deviation

	①		②		③	
	Auditing Unavailable		Auditing Available		Auditing Available	
α	0.833 (5/6)		0.833 (5/6)		0.833 (5/6)	
β			0.8 (4/5)		0.8 (4/5)	
C			1		1.5	
N	28		28		28	
Standard Deviation of Audit Purchase Frequency			3.179		3.336	
Average of Audit Purchase Frequency			16.57		16.36	
	Effort	Investment	Effort	Investment	Effort	Investment
SD	4.573	5.715	3.903	3.454	3.276	2.81
Average	6.893	9	14.14	13.82	12.71	12.25
t-Test	① and ②		① and ③		② and ③	
t-Statistic (T)	6.380	3.820	5.476	2.700	1.483	1.867
p-Value (p)	**0.0000	**0.0002	**0.0000	**0.0046	0.0719	*0.0336
F-Statistic (T)	1.372	2.738	1.949	4.136	1.420	1.510
p-Value (p)	0.416	**0.0110	0.0887	**0.0004	0.3678	0.2901
Welch-Statistic (T)	6.380	3.820	5.476	2.700	1.483	1.867
p-Value (p)	**0.0000	**0.0002	**0.0000	**0.0051	0.0720	0.0337
Kruskal-Wallis Test	① and ② and ③					
H-Statistic (H)	29.17	12.52				
p-Value (p)	**0.0000	**0.0019				
Mann-Whitney U's test					② and ③	
Z-Statistic(Z)					-1.442	-1.816
p-Value(p)					0.0747	*0.0341
Effort Selection and Investment	0.7978		0.7217		0.6679	
Effort Selection and Audit Purchase			0.3185		0.2063	
Audit Purchase and Investment			0.5257		0.4405	

H2: The number of investment is higher when auditing available.
H3: The number of effort made by managers is higher at the high audit fee.
H4: The number of investment is higher at the high audit fee.

The comparison and the statistical examination are made between the three markets (Table1). The first two hypotheses receive very strong support, whereas the last two don't, since the number of effort and investment decreases significantly at the high audit fee. The digression analysis yields less convincing results. The coefficiency between the frequency of effort and investment is lower when auditing available and lowest at the high audit fee.

The very low co-efficiency between the frequency of effort and audit purchase shows that managers try to cheat the investors. The low coefficiency between the frequency of audit purchase and investment indicates investor's defection from the market.

The statistical analysis of the players payoffs shows almost same results (Table2). The payoff of managers is highest when auditing unavailable but almost same as when auditing available at the low audit fee. Their payoff decreases significantly when auditing available at the high audit fee. It is partially due to the increasing cost of audit purchase. Moreover, investor's defections pushed by the fear of being cheated can play a very important role. The payoff of investors is minus when auditing unavailable. It indicates that the investors get hurt by cheating. Their payoff increases considerably when auditing available at the low audit fee. Its impressive display demonstrates the economic efficiency achieved by auditing, but their payoff decreases significantly at the high audit fee. Taking a look at the selection of effort and investment over time (Figure 1 and 2), the frequency is always much higher when auditing available but slightly downward over time.

Table 2. Results (Payoffs), ** Significantly Different at $p = 0.01$, *Significantly Different at $p = 0.05$ (both tails)

	①		②		③	
	Auditing Unavailable		Auditing Available		Auditing Available	
α	0.833 (5/6)		0.833 (5/6)		0.833 (5/6)	
β			0.8 (4/5)		0.8 (4/5)	
C			1		1.5	
N	28		28		28	
Payoff (yen)	Manager	Investor	Manager	Investor	Manager	Investor
Standard Deviation	41.72	15.21	23.20	19.11	19.45	15.74
Average	55.89	-0.357	51.07	53.21	34.39	41.42
t-Test	① and ②		① and ③		② and (3)	
	Manager	Investor	Manager	Investor	Manager	Investor
t-Statistic (T)	0.5344	11.61	2.471	10.100	2.915	2.519
p-Value (p)	0.5953	**0.0000	*0.0167	**0.0000	**0.0052	**0.0148
F-Statistic (T)	3.234	1.579	4.603	1.071	1.423	1.473
p-Value (p)	**0.0033	0.2419	**0.0002	0.8590	0.3646	0.3200
Welch-Statistic (T)	0.5344	11.61	2.471	10.100	2.915	2.519
p-Value (p)	0.5953	**0.0000	*0.0170	**0.0000	**0.0052	**0.0149
Kruskal-Wallis Test	① and ② and ③					
H-Statistic (H)	8.857	53.07				
p-Value (p)	**0.0119	**0.0000				
Mann-Whitney U's Test					② and ③	
Z-Statistic(Z)					2.917	2.451
p-Value(p)					**0.0035	*0.0142

4 Concluding Remarks

At the high audit fee managers find too high the total incremental costs of inducing trust on the part of the investors. They thus want to spare the cost of effort or audit purchase. Facing with this alternative, what they have chosen is to cut down on the former, more essential one. That causes investor's defections. If the players find the total signaling costs too high, they are rather likely to spite the others.

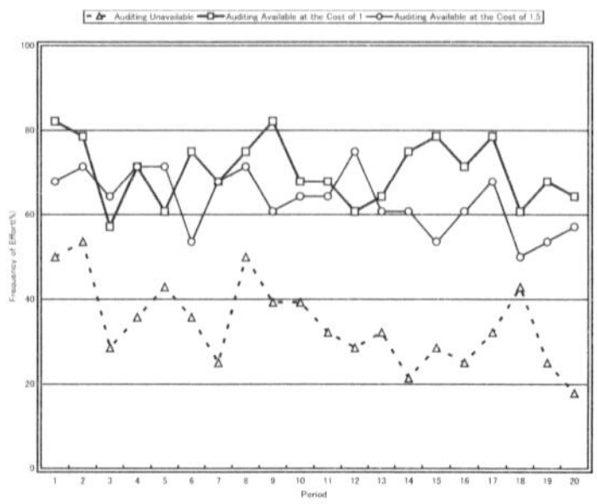

Fig. 1. Selection of Effort over Time

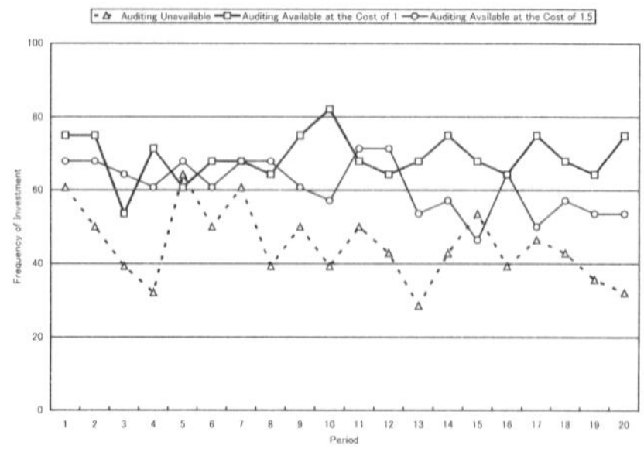

Fig. 2. Selection of Investment over Time

References

1. Dejong, D.V., R. Forsythe, R.J. Lundholm, and W.C. Uecker, A Laboratory Investigation of the Moral Hazard Problem in an Agency Relationship, Journal of Accounting Research, Vol.23 Supplement, (1985), pp.81-120.
2. Dopuch, N. and R. R. King, Negligence Versus Strict Liability in Auditing: An Experimental Investigation, The Accounting Review, Vol.67, No.1, (1992), pp. 97-120.
3. Dopuch, N, R.R. King, and D.E. Wallin, The Use of Experimental Markets in Auditing Research: Some Initial Finding, Auditing: A Journal of Practice & Theory, Vol.8, Supplement (1989), pp.98-136.
4. Kachelmeier, S.J., A Laboratory Market Investigation of the Demand for Strategic Auditing, Auditing: A journal of Practice & Theory, Vol.10, Supplement (1991), pp.25-53.
5. Kato, T., Audit Credibility: A Theory and an Experimental Investigation, Bulletin of Faculty of Commerce Meiji University, Vol.86, No4, (2004), pp.21-37.
6. Mayew, B.W., J.W. Schatzberg, and G.R. Sevcik, The Effect of Accounting Uncertainty and Auditor Reputation on Auditor Objectivity, Auditing: A Journal of Practice and Theory, Vol.21, No.1, (2001), pp. 49-70.
7. Mayew, B.W. and J.E. Pike, Does Investor Selection of Auditors Enhance Auditor Independence? The Accounting Review, Vol.79, No.3, (2004), pp.797-822.
8. Wallin D.E., Legal Recourse and the Demand for Auditing, The Accounting Review, Vol.67, No.1, (1992), pp. 121-147.

When Firms Contest in Markets: An Experiment *

Utteeyo Dasgupta

University of Arizona

1 Introduction

Regulating a natural monopoly market has always remained a source of concern. The problem arises because of the decreasing average cost structure in the market. Ideally, only one firm serving the whole market demand is the efficient solution to avoid any cost duplication. However, when there is a single unregulated firm serving a market it brings up the standard monopoly price-gouging problem. Many utility services share the characteristics of a natural monopoly. As a result, almost all countries in their deregulation phases are concerned with the efficient running of such markets. Restraining monopoly behavior effectively in a natural monopoly market remains a much-debated issue. The idea of creating a "contestable" [1] environment has influenced USA, UK and many other countries during their deregulation phase. In a perfectly contestable market [2] the threat of hit-and-run entry by new entrants in the monopoly market can provide the right disciplining stick for the monopolist incumbent to charge a price equal to the average cost of production (the Ramsey optimal price). This outcome is described as a contestable (market) outcome [3] .

* I thank James C. Cox, Martin Dufwenberg, Stanley Reynolds the participants at the ESA 2004, EES 2004 conferences and the editorial board for useful and critical sugestions. I also extend my gratitude to Professor Hidetoshi Yamaji and the organizers of the EES 2004 conference for giving me an opportunity to present my work. A special thanks to Todd Sorenson for his insight in programming the software. Financial Support from the Experiment Science Laboratory at the University of Arizona is gratefully acknowledged.

[1] William Baumol in his 1981 address as outgoing president of the American Economic Association put forward the idea of a contestable market.

[2] A perfectly contestable market is devoid of any sunk entry costs. So an entrant can enter a market whenever there are profit opportunities without worrying about sunk costs.

[3] Baumol Panzar and Willig [1]

Although the role of an entrant firm "waiting on the wings" is crucial to the contestable outcome, the literature [4] on contestability does not explicitly model the origin or nature of entrant firms. Who are the possible entrants in the natural monopoly markets? Casual empiricism suggests that the possible entrants are typically firms offering similar products in other markets. For example, in airlines Southwest and United Airlines have their own airport dominance and are possible entrants in each other's routes. In high-speed DSL services Qwest, Verizon and Earthlink are firms who have their own home markets but are the potential entrants in each other's markets. We are interested in the following question: Does the threat of hit and run entry invoke the right disciplining effect when entrants have their own home markets? If entrant firms have their own home markets it is plausible that although another market is vulnerable to "hit-and-run" entry, the entrant might be reluctant to enter if it fears retaliation in its own market. A "hit-and-run" strategy might not work so well when firms have their own monopoly markets and have opportunities to enter each other's markets. The point to note is that although the entrant can 'hit', it cannot quite 'run' away because the incumbent might decide to enter and price aggressively in the entrant's own market in an effort to punish him. Since the contestable outcome relies on the possible threats of entry, a possible price war in each other's market can mitigate that threat to a large extent. This in turn might make the firms behave in more collusive manners.

In this paper I evaluate for the first time the role of contestability in an experimental framework where each firm has its own home market and are potential entrants in each other's markets [5] . My experimental results establish that when each firm has its own monopoly market then contestable outcome is not observed. Instead prices in the monopoly markets are close to the theoretical monopoly prices. In contrast, if the entrant firm comes from a competitive environment then we do observe the threat of entry to provide enough disciplining force in the pricing behavior of the monopoly incumbent. Hit-and-run entry in initial periods makes the monopolist incumbent price at average cost in later periods.

---

[4] See Baumol Panzar and Willig [1], Brock [3] and Schwartz [16] for a discussion on the assumptions on contestability.

[5] Earlier experimental work by Coursey, Issac and Smith [6], Harrison and Mckee [12], Harrison, Mckee and Rutstrom [13] had the following design: There is a single contested market. At least two firms choose price and quantity simultaneously like in a Bertrand market. The outcomes conform to the contestable outcome. In an alternative design mechanism the subjects could buy or sell futures in a double-auction market among them to sell or buy them in a spot market with predetermined (but unknown) prices(I thank the editorial board for pointing this out). Harrison [10] modified the above design to introduce sequential moves by the incumbent and the entrant and found even stronger support for the contestable outcomes.

2 Experimental Design

We consider a simple two-firm two-market setup in our experimental design. Each firm is an incumbent in one of the markets and has the option to enter the other market [6]. Our baseline treatment consists of two natural monopoly markets with the same demand and cost conditions. We will call this the monopoly treatment. In our contrasting treatment one of the markets is a natural monopoly market and the other market is a stylized perfectly competitive market. The demand and cost conditions in the monopoly market are the same as the baseline treatment. The payoff in the competitive market is fixed at 1 experimental dollar and there are no costs. This payoff is comparable to the payoff the incumbent gets in the natural monopoly market when it charges a price equal to the average cost of production (Ramsey price). The payoff in the competitive market remains unaffected by entry. We will refer to this as the perfect competition treatment [7].

The experiment is setup as a two-stage game. In the baseline treatment each firm chooses a price in its home market in the first stage. In the second stage the firm gets to know the 1st stage price in the other market and decides whether or not to enter and choose a price in the other market. In the contrasting treatment, the incumbent firm chooses a price in the monopoly market in the first stage. In the second stage, the competitive entrant gets to know the 1st stage price in the other market and decides whether or not to enter and choose a price in the other market. The profits are realized at the end of the two stages, where the firm with the lower posted price in the monopoly market realizes profits in that market. The two-stage decision operationalizes the fact that each firm can actually evaluate the profitability in the other market before it decides whether to enter the market [8] The natural monopoly market has the following demand and cost functions:

$$p = 20 - q$$
$$c(q) = 32 + 2q$$

where p is the price and q is the quantity demanded, and c(q) is the cost function. A firm only incurs costs if it sells in the monopoly market. This quasi-fixed nature of the cost operationalizes the costless exit from a market, an assumption crucial for a perfectly contestable market. In all the treatments, the sellers interact repeatedly with the same person with a random stopping

[6] Dasgupta [7] analyses a model for possible entrant and incumbent firm behaviors in such a context.

[7] Dasgupta [8] analyses a Bertrand price matching environment as an alternative competitive market environment as well.

[8] This is one of the critical behavioral assumptions that contestability hinges on and has been absent in almost all of the earlier experiments on Contestability (See Harrison [10] for a discussion).

rule to avoid any end game effect. The number of firms in the market and the exchange rate of experimental earnings into real currency are clearly explained in the instructions to the participants [9].

The experiments were conducted in the Economic Science Laboratory (ESL) with the undergraduates at the University of Arizona. The experiment was programmed and conducted with the software z-Tree (Fischbacher [19]). Table 1 summarizes the number of sessions run and the profit levels in the two treatments.

Table 1. Design and Sessions

Treatment	Number of Firms	Profits in the Monopoly market		Profit in the Treatment market		Sessions Run
		Max	Min	Max	Min	
Monopoly	2	50	1	50	1	7
Perfect Competition	2	50	1	1	1	7

3 Experiment Results

We hypothesize that an entrant from a competitive market disciplines a monopoly incumbent in a perfectly contestable market. In contrast, an entrant with its own monopoly market open to entry tacitly colludes with the monopoly incumbent of the other market.

Our main variable of interest is the average prices over periods in the natural monopoly market for each of the treatments. Using a Mann-Whitney test we find the average price in the monopoly treatment significantly higher than in the perfect competition treatment (one-tailed test) [10]. Figure 1, compares the effects the two treatments have on the average prices in the monopoly market. The first 5 periods are not reported since the subjects are learning about decisions in the market although the choices aren't starkly different.

We define a measure to illustrate the disciplining effects of contestability in the two treatments. Let C be a contestability index where

$$C = (\Pi_m - \Pi_a)/(\Pi_m - \Pi_0)$$

Where Π_a is the actual profit observed in the experimental monopoly market, Π_m is the theoretical monopoly profits in the market and Π_0 is the profit associated with the Ramsey pricing. If $\Pi_a = \Pi_m$, then $C = 0$ and contestability fails to discipline monopoly behavior. Instead when $C = 1$ then contestability is successful in disciplining the monopoly market. Figure 2 illustrates the average value of C across all periods in the monopoly market for each treatment.

[9] Subject instruction sets can be obtained by a request to the author.

[10] See Dasgupta [8] for a comparison of prices in the Bertrand vs. Monopoly treatment

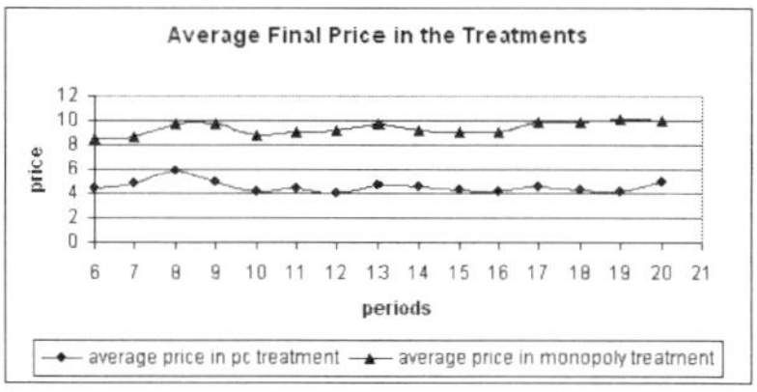

Fig. 1. Comparison of Average Prices (monopoly and perfect competition treatment)

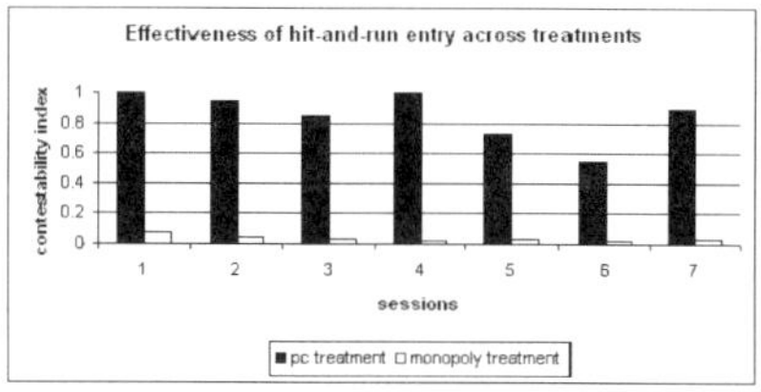

Fig. 2. Contestability index in the two treatments

The difference in C across treatments is quite stark. While the monopoly treatment yielded C close to zero in each of the markets, the other two treatments had C close to one or actually one.

4 Conclusion

In this experiment we tested for the effectiveness of hit-and-run entry in perfectly contestable markets when the entrant and the incumbent each have their own home markets. We find that the threat of entry is successful in lowering prices only if the entrant has relatively lower profits in its own market as compared to the monopoly market, or the entrant's profits are unaffected by retalliatory entry. In the perfect competition treatment prices in the natural monopoly market quickly adjusted to the entry restricting average cost level over periods. However, in the monopoly treatment prices remain very close to the theoretical monopoly price levels.

My results demonstrate that contestability theory in its conceptualization might have overlooked an important dimension in firm interactions. In particular the entrant's market, and the interaction environment between the firms need to be incorporated for its successful implementation. The threat of entry is indeed a potential disciplining force when the entrant has very little to lose in the long run by entering an incumbent's market. On the other hand an entrant from a monopoly market would be wary of vigorous price competition in the other monopoly market because in the long run it might depress profits in all markets [11] .

References

1. Baumol W, Panzar J. and Willig R(1982) Contestable Markets and the Theory of Industry Structure Harcourt Brace Jovanovich, New York.

[11] Bernheim and Whinston [2] suggest similar mutual forbearance results in duopoly markets in a more general framework.

2. Bernheim D and Whinston M. (2000) Multimarket Contact and Collusive Behavior Readings for Contemporary Economics. Malden, Mass. And Oxford Blackwell
3. Brock W. (1983) Contestable Markets and the Theory of Industry Structure: A Review Article Journal of Political Economy 91: 1055-56
4. Brown-Kruse Jamie L. (1991) Contestability in the Presence of an Alternate Market: AN Experimental Examination The RAND Journal of Economics 22: 136-147
5. Call Gregory D., Keeler Theodore E. (1985) Airline Deregulation, Fares, and Market Behavior: Some Empirical Evidence In: Daughety Andrew F (Eds) Analytical Studies in Transport Economics Cambridge: Cambridge University Press: 221-47
6. Coursey D, Issac R, and Smith V. (1984) Natural Monopoly and Contested Markets: Some Experimental Results Journal of Law and Economics 27: 91-113
7. Dasgupta Utteeyo (2005) Contesting Contestability Working Paper
8. Dasgupta Utteeyo (2005) Minding Your Own Business Working Paper
9. Dixit A (1980) The Role of Investment in Entry Deterrence Economic Journal 90: 95-106
10. Evendyn Rhys D, Williams Alan W. (2000) Contestability: The Debate and Industry Policy Economic Analysis and Policy v30: 75-91
11. Harrison G. (1987) Experimental Evaluation of the Contestable Markets Hypothesis. In: Bailey Elizabeth (Eds) Public Regulation The MIT Press
12. Harrison G, McKee M. (1985) Monopoly Behavior, Decentralized Regulation, and Contestable Markets: An Experimental Evaluation Rand Journal of Economics 16: 51-69
13. Harrison G, McKee M. and Rutstrom E.E. (1987) Experimental evaluations of institutions of monopoly restraints :In Green L and Kagel J. (Eds) Advances in Behavioral Economics, vol. 2 Norwood, N.J: Ablex Press, 54-94
14. Rapoport Amnon, Seale Darryl A. and Winter Eyal (2000) An experimental study of coordination and learning in iterated two-market entry games Economic Theory 16 : 661-687
15. Schwartz M. and Reynolds R (1983) Contestable Markets: An Uprising in the Theory of Industry Structure: Comment American Economic Review 73: 488-90
16. Schwartz M. (1986) The Nature and Scope of Contestability Theory Oxford Economic Papers, New Series, Vol. 38, Supplement: Strategic Behavior and Industrial Competition : 37-57
17. Shepherd W (1984) "Contestability" vs. "Competition," American Economic Review 74: 572-87
18. Tye W (1985) On the Applicability of the theory of Contestable Markets to Rail/Water Carrier Mergers Logistics and Transportation Review 21: 57-76
19. Urs Fischbacher: z-Tree - Zurich Toolbox for Readymade Economic Experiments - Experimenter's Manual, Working Paper Nr. 21, Institute for Empirical Research in Economics, University of Zurich, 1999
20. Weitzman M. (1983) Contestable Markets: An uprising in the Theory of Industry Structure: Comment. American Economic Review 73: 486-487
21. Zwick Rami, Rappoport Amnon (2002) Tacit Coordination in a Decentralized Market Entry Game with Fixed Capacity Experimental Economics 5: 253-272

How to Use Private Information in a Multi-person Zero-sum Game *

Hiroyasu Yoneda[1], Gen Masumoto[2], and Sobei H. Oda[3]

[1] Kyoto Sangyo University
[2] Kyoto Sangyo University
[3] Kyoto Sangyo University

1 Introduction

This paper describes how people play a zero sum game with different private information. Apparently more informed players earn more than less informed players do. What happens however if people buy and sell speculatively in the future market? Those who are better informed seem to have greater chance to earn money, while those who have no information may expect zero profit because they seem to have equal chance to make money (to buy a commodity whose price will increase or to sell a commodity whose price will decrease) and to lose money (to sell a commodity whose price will increase or to buy a commodity whose price will decrease) . Yet the sum of all traders is zero. If the most informed player earns profit and the lest informed player expects zero profit, some modestly informed players must suffer loss.

Huber and Kirchler [1] formulated a simple model of the future market and observed the paradoxical distribution of gain and loss among players in their simulations with computer agents and experiments with human subjects. In this paper we shall develop their analysis to examine the strategies of players (agents and subjects) and income distribution among them in more detail.

This paper is organised in the following way. In the next section we shall define our model. In Section 3 we shall show some results of our simulations: income distribution changes drastically if some or all agents change their strategies. In Section 4 we shall present some findings of our experiments: human subjects change their strategies so flexibly to increase their profit according to the strategies of their competitors.

---

* This paper is a shorter version of Yoneda, Masumoto and Oda [2] with some new findings in simulations and experiments but without details in mathematical analysis. The present study is based on the research by the Open Research Centre Project "Experimental Economics: A new method of teaching economics and the research on its impact on society" and Grants-in-aids for Scientific Research 17310029.

2 The Game

The game is played in the following order by $2M+1$ players.[1]

1. At the beginning of each round Player i ($0 \leq i \leq 2M$) tells his reservation price R_i to the auctioneer to make the contract that he will buy a unit of a future commodity if R_i is higher than its price P and that he will sell a unit of the commodity if $R_i < P$.
2. The auctioneer gathers all the reservation prices and declares the median of them as P so that demand equals supply in the future market.
3. The true value of the commodity V is determined exogenously as the sum of $2M$ stochastic variables: $V = \sum_{k=1}^{2M} X_k$ where X_1, X_2, ..., and X_{2M} are determined identically and independently to be 0 or 1 with equal probability 0.5.
4. Those who bought (sold) the commodity in the future market sells (buys) it in the spot market at the true value V to close their accounts. Hence each player's profit is determined as soon as V is revealed.

Every player's profit is determined as a result of the above-mentioned trade: if $P < R_i$, Player i buys the commodity at P in the future market and sells it at V in the spot market to earn $V - P$ of profit; if $R_i < P$, Player i sells the commodity at P in the future market and buys it at V in the spot market to earn $P - V$ of profit. Needless to say, the sum of all player's profit is always zero: $\sum_{i:V<R_i}(P-V) + \sum_{i:R_i<V}(V-P) = 0$.

3 Simulations With Computer Agents

Let us assume that all players know how V is determined: $V = \sum_{k=1}^{2M} X_k$ and that before determining R_i, Player i can correctly see what values X_1, X_2, ..., and X_i will be (Player $2M$ can predict the values of all the $2M$ variables while Player 0 can forecast nothing). In the circumstances Player i can take it into account to determine R_i that V will be between $V_i^{\min} = \sum_{k=1}^{i} x_k$ and $V_i^{\max} = \sum_{k=1}^{i} x_k + (2M - i)$. As is readily checked, a strategy that may choose a value smaller than $V_i^{\min}$ or greater than $V_i^{\max}$ as R_i is weakly dominated by such a strategy that chooses $R_i = V_i^{\min}$ or $R_i = V_i^{\max}$ then.

On the above-mentioned assumption and consideration, we run a number of simulations with 101 agents (namely $M = 50$) which follow either **the middle-value strategy**: $R_i = \frac{1}{2}(V_i^{\min} + V_i^{\max})$, or **the either-end strategy**: $R_i = V_i^{\min}$ or $R_i = V_i^{\max}$ with equal probability.

How gain and loss are distributed among the agents are illustrated in Figures 1-4. There the horizontal axis stands for the agent (from 0 to 100) while the vertical axis represents each player's average profit (for 10000 rounds).

[1] The game is as the same as the one presented by Huber and Kirchler (2004) except that the number of players is not even but odd.

The marginal contribution of private information to profit, which is defined as Player $i+1$'s profit minus Player i's, is not monotonously increasing (Figures 1 and 2). An agent may decrease his loss (and increase his gain) by adopting the strategy which is different from the strategy all the other agents follow (Figures 3 and 4).[2]

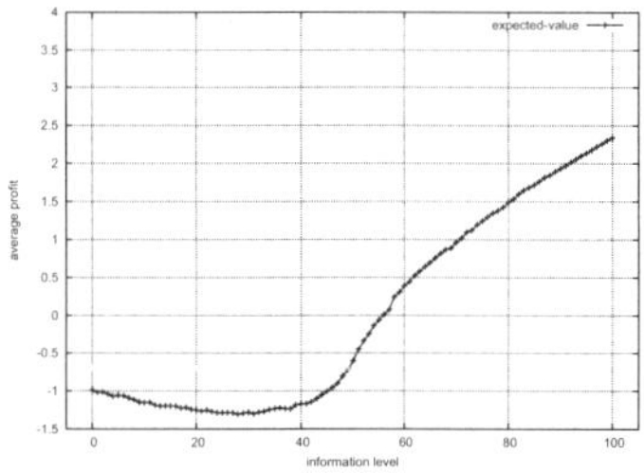

Fig. 1. Income distribution among middle-value strategy agents

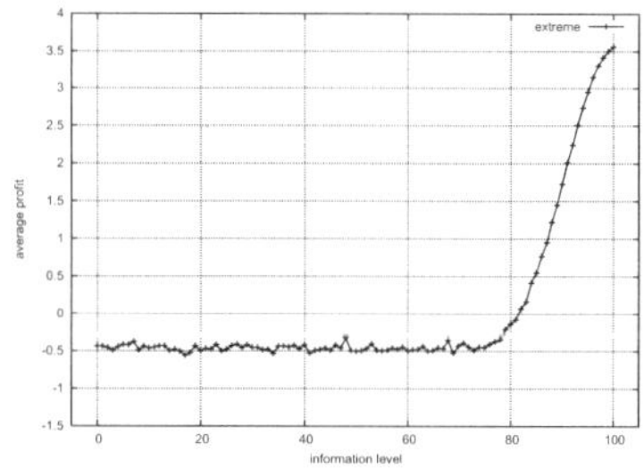

Fig. 2. Income distribution among either-end strategy agents

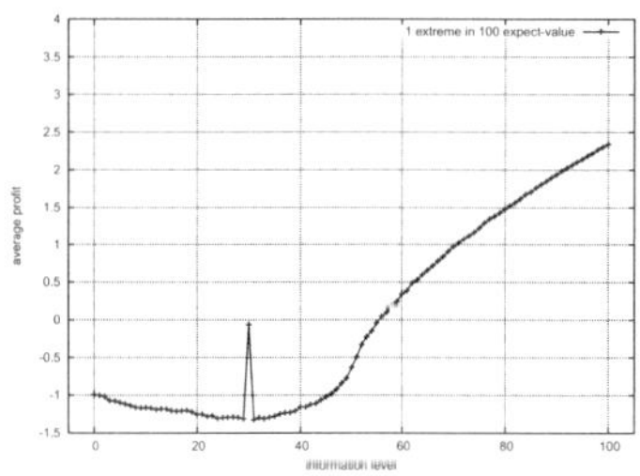

Fig. 3. Income distribution among an either-end strategy agent (Player 30) and middle-value strategy agents

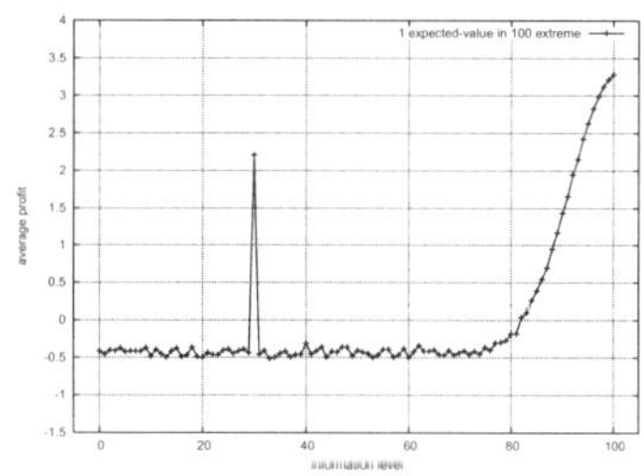

Fig. 4. Income distribution among a middle-value strategy agent (Player 30) and either-end strategy agents

Figure 5 shows how Player 30's profit changes if he follows the following strategy.

$$R_i \begin{cases} = V_i - \theta_i(V_i - V_i^{\min}) & \text{with probability } 0.5 \\ = V_i + \theta_i(V_i^{\max} - V_i) & \text{with probability } 0.5 \end{cases}$$

where $i = 30$, $0 \leq \theta_i \leq 1$ and $V_i = \frac{1}{2}(V_i^{\min} + V_i^{\max}) = \sum_{k=1}^{i} x_k + \frac{1}{2}(2M - i)$ (the strategy is the middle-value strategy if $\theta_i = 0$, while it is the either-end

[2] Except for small differences in the number of agents and their strategies, Figures 1 and 3 were discovered by Huber and Kirchler (2004).

strategy if $\theta_i = 1$). Player 30's profit increases as θ_{30} increases if all the other players follow the middle-value strategy (the solid curve), while it decreases as θ_{30} increases if all the other players follow the either-end strategy (the broken curve).[3]

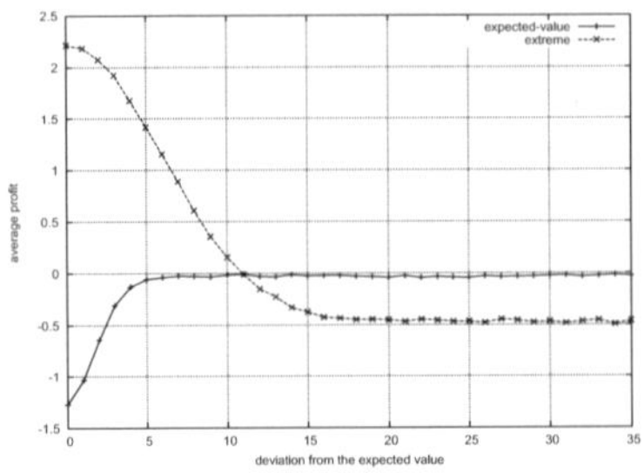

Fig. 5. The effect of $|R_{30} - V_{30}|$ (=$35\theta_{30}$) on Player 30's income

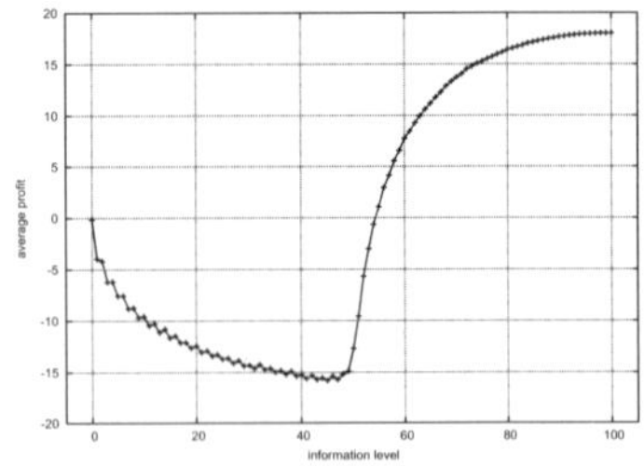

Fig. 6. Income distribution among deterministic either-end strategy agents

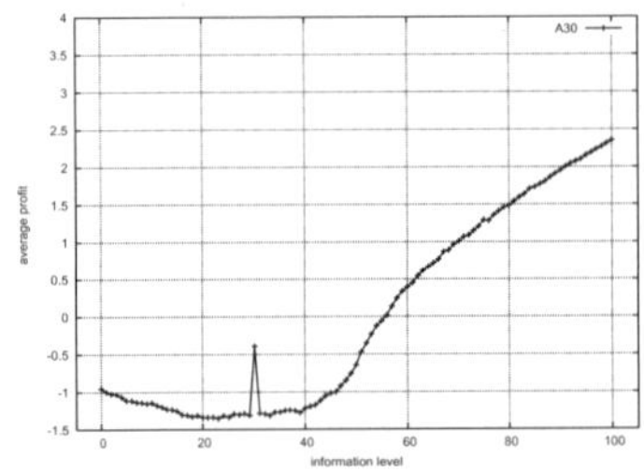

Fig. 7. Income distribution among a human subject (Player 30) and middle-value strategy agents

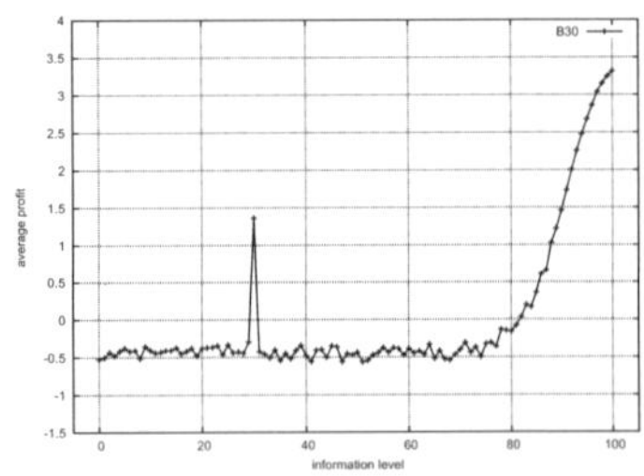

Fig. 8. Income distribution among a human subject (Player 30) and either-end strategy agents

[3] Though omitting mathematical analysis, we should only mention the following to illustrate how the same strategy produces different profit if others change their strategies. It is not a coincident that Player 30's profit is zero in Figure 3 and he can earn positive profit if he chooses $V_i^{\min}$ or $V_i^{\max}$ not randomly but systematically: $R_i = V_i^{\min}$ if $V_i < \frac{1}{2}i$; $R_i = V_i^{\max}$ otherwise. Yet income distribution in Figure 6, which is realised if all players adopt the above-mentioned strategy, is more uneven than it is in Figures 1 and 2.

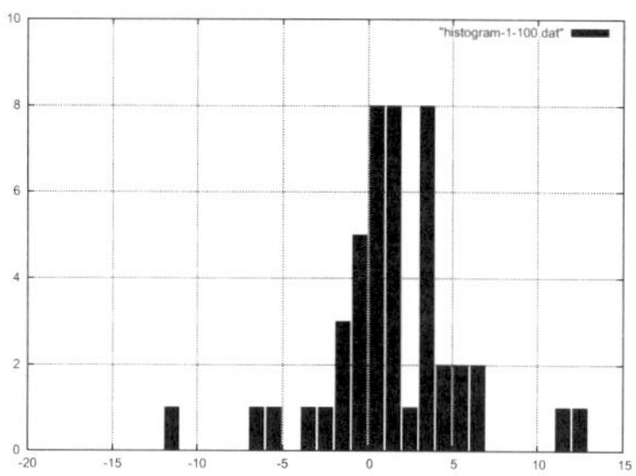

Fig. 9. Increase in $|R_{30} - V_{30}|$ from Session (b) to Session (c)

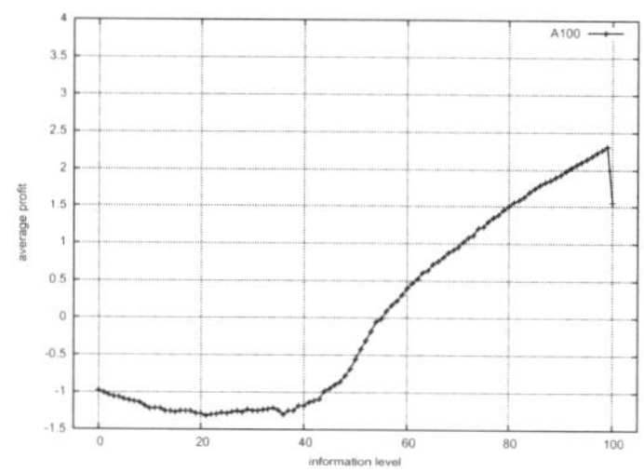

Fig. 10. Income distribution among a human subject (Player 100) and middle-value strategy agents

4 Experiments With Agents and Human Subjects

We did experiments with computer agents and human subjects at Kyoto Experimental Economics Laboratory (KEEL), Kyoto Sangyo University (KSU) on October 13 and 16, 2004. In total 46 undergraduates of KSU played the game mentioned in the previous section as a unique human player with 100 computer agents. To put it concretely, each subject played the game (a) with 100 middle-value strategy agents for 100 rounds as Player 100; (b) with 100 middle-value strategy agents for 100 rounds as Player 30; (c) with 100 either-end strategy for 100 rounds as Player 30. Twenty one subjects played the three sessions in the above-mentioned order, while the other subjects played the last two sessions reversely.

Figures 7 and 8 show how gain and loss were distributed among human subjects and computer agents in Sessions (b) and (c) respectively. There the horizontal and vertical axes are the same as in the previous figures; the data in Figure 7 is aggregated from Session (b) played as the second session and Session (b) as the third session, because the performance of players (subjects and agents) are not different between the two sessions; a similar remark applies to the data in Figure 8.

Although it is not as large as it is in Figures 3 and 4, an decrease in loss or increase in gain of Player 30 is visible in Figures 7 and 8, which fact suggests that our subjects changed their strategies according to the strategy of their competitors. It is confirmed by Figure 9, where about seventy per cent of our subjects chose R_{30} in more distance from V_{30} in (b) than they did in (c) (Wilcoxon signed rank test, z=2.873, P=0.004, two-tailed). Remembering Figure 5, we can see it increased their profits.

5 Concluding Remarks

The results of our experiments may suggest that human players can think and/or learn so flexibly that they can change their strategies according to changes in their circumstances. With partial information and ignorance of their competitors' strategy, most human subjects outwitted their competitors (computer agents). This performance is even more impressive if we take it into account that not a few subjects failed to find out the best strategy ($R_{100} = V_{100}$) when they have the full information (Figure 10).

However, the good performance of our subjects may be benefited largely from the fact that their rivals are all such simple computer agents that cannot change their strategies. If they also could change their strategies according to their experience, the dynamics of the game would be so complicated that they could not be outwitted by human subjects.

References

1. Jüergen Huber and Michael Kirchler (2004): "The Value of Information in Markets with Heterogeneously Informed Traders - and Experimental and a Simulation Approach", presented at the 9th Workshop on Economics and Heterogeneous Interacting Agents, Kyoto University, Kyoto, Japan, 27-29 May 2004.
2. Yoneda, H., Masumoto, G. and Oda, S. H. (2004): "Marginal Contribution of Information to Profit in a Zero-sum Game", presented at Experiments in Economic Sciences: New Approaches to Solving Real-world Problems, Okayama International Hotel and Kyoto Sangyo University, Okayama and Kyoto, Japan, 14-17 December 2004; in *The proceedings of the Experiments in Economic Sciences: New Approaches to Solving Real-world Problems* <http://www.kyoto-su.ac.jp/project/orc/execo/EES2004/proceedings.html>, pp.770-786.

A Price Competition Experiment Between Middlemen: Linear Function Case *

Kazuhito Ogawa[1], Kouhei Iyori[2], and Sobei H. Oda[3]

[1] Hiroshima City University
[2] Hiroshima Shudo University
[3] Kyoto Sangyo University

1 Introduction

A middleman is important; he/she purchases goods from suppliers for resale, creates and manages markets, seeks out suppliers, finds and encourages buyers, selects bid and ask prices, and holds inventories to provide liquidity or make services and goods available. However, economic theories such as the general equilibrium theory have not examined the role of a middleman.

The market microstructure theory, which emerged in the 1980s, takes the role of a middleman into account. Spulber [7] focuses on how price-setting middlemen offer prices and compete each other, reduce uncertainty, and help buyers and sellers to meet and carry out transactions. The models assume that a middleman sets a bid price and buys a commodity from a supplier, then sells it to a buyer at a higher ask price.

We investigates whether two middlemen offer competitive prices as bid and ask prices in a laboratory setting. Subsequently, we examine how the information about supply and demand and the experience of the subject affect their behavior. Finally we compare our results with other experimental results.

2 Theory and Experimental Setting

The following is an explanation of the model proposed by Stahl [8] and Spulber [7]. Middlemen, who are the subjects of the experiment, offer prices sequentially, competing first for supplies and then for customers. The inputs and outputs are homogeneous.

* This research was supported by the Open Research Centre "Experimental Economics: A new method of teaching economics and the research on its impact on society, "Kyoto Sangyo University, Japan Society for the Promotion of Science(JSPS), Grant-in-Aid for Scientific Research (B), 13480115, and JSPS Grant-in-Aid for Doctoral Candidate Fellows, FY2002-FY2004.

In the first stage, the middlemen offer bid prices simultaneously. A simulated seller sells an input if the price offered is higher than her production cost. If the middlemen offer identical prices, each obtains half of the available supplies at that price. If they offer different prices, the highest-priced firm obtains all the available supplies. The other firm leaves the market.

$$q_{Si} = \begin{cases} 0 \text{ if } w_i < w_j \\ S(w_i)/2 \text{ if } w_i = w_j \\ S(w_i) \text{ if } w_i > w_j, \end{cases} \tag{1}$$

where q_{Si} and w_i denote respectively the number of supply units and bid price of player i.

In the second stage, only the middleman who has obtained the commodity sells it to customers. A simulated customer buys the commodity if the offered price is lower than her reservation value. If two middlemen have the commodity, they offer ask prices simultaneously. If they offer different prices, the lowest-priced middleman sells first. Customers with the low reservation prices buy first in order to trade as much of the commodity as possible. If they offer identical prices, each sells half of the available demands at that price.

$$q_{Di} = \begin{cases} 0 \text{ if } q_{Si} = 0 \\ D(p_i) \text{ if } p_i < p_j, \text{ and } q_{Si} \geq D(p_i) \text{ or if } q_{Sj} = 0, \text{ and } q_{Si} \geq D(p_i) \\ q_{Si} \text{ if } p_i < p_j, \text{ and } q_{Si} < D(p_i) \text{ or if } q_{Sj} = 0, \text{ and } q_{Si} < D(p_i) \\ q_{Si} \text{ if } p_i > p_j \text{ and } q_{Si} < D(p_j) - q_{Sj} \\ D(p_j) - q_{Sj} \text{ if } p_i > p_j \text{ and } q_{Si} > D(p_j) - q_{Sj}. \end{cases} \tag{2}$$

Finally, player i's profit is $\pi_i = p_i q_{Di} - w_i q_{Si}$.

If $p^R \leq p^W$, this game has a unique sub-game perfect Nash equilibrium (SPNE), where p^R and p^W indicate the sales-revenue-maximizing ask price and the Walrasian price, respectively. In SPNE, both the middlemen set the competitive price as the bid and ask prices.[1]

The supply ($S(w) = 2w$) and the demand ($D(p) = 104 - 2p$) functions were used under the Liner Function Treatment (LFT).[2] These functions satisfy $p^R \leq p^W$. Due to the discontinuity, the SPNE is (27, 25), where (a, b) indicates ask and bid prices. The zero-profit Bertrand outcome is the Nash equilibrium (26, 26).[3] Since in terms of economic intuition these equilibria are much the same, we regard (27, 25) and (26, 26) as competitive equilibria. The profit-maximizing price set in the monopoly case is $(39, 13)$.

The LFT comprised three treatments. Under T-1, the inexperienced subjects were not aware of the detail of supply or demand functions. Under T-2, subjects who had already experienced T-1 or T-3 were aware of the detail of these functions. Under T-3, the inexperienced subjects were aware of the details of these functions.

[1] The detailed proof can be obtained from Spulber [7], ch. 3 pp. 65-66 and Stahl [8].
[2] In fact, Ogawa et al. [4], [5] conducted the Step-Function Treatments (SFT).
[3] We do not include (26, 25) in the competitive price set since it is not a rational alternative in terms of the best response.

The experiment was conducted at the Kyoto Sangyo University Experimental Economics Laboratory (KEEL) between October 2002 and April 2004. A total of 114 subjects participated in the LFT. In each session, all the students entered the KEEL, received instructions, and were told that they would receive 2000 JPY for showing up (approximately 16 dollars at the time of the experiment) and additional monetary rewards contingent on the total performance in the experiment.[4] The contingent part was $0.07 * \sum$ total profit.

Each of the treatments had four features. First, there were more than 80 rounds, and the subjects were not informed about the number of rounds. Second, the subjects played the game with an identical opponent but were not aware of who the opponent was. Third, in every round, they were informed about their bid prices and inputs, and the opponent's bid and ask prices, their ask price, their sales and profit. Finally they played the ten-round training session. The opponents were different from the ones under T-i.

3 Results

The following is an overview of the results. The average profit of the subjects per round is the highest under T-2. The difference is significant at the 5 % level for all cases (Table 1). The maximum total profit and the minimum total profit under T-2 is the highest among all the treatments.

The profit difference is mostly duo to the burden of the dead inventory: $q_S - q_D$. Since the subjects under T-1 were not aware of the supply or demand, they sought the profitable price sets first. While doing this, they often suffered from a dead inventory; however, the subjects under T-2 and T-3 possessed the knowledge of the functions in advance and could choose prices in such a way that they did not have to face a dead inventory. Moreover, since the subjects under T-2 were experienced, they earned more profit.

Table 2 represents the average winning prices in the last 30 rounds. Applying the F-test and t-test to compare the result under T-1 with that under T-3, the bid-ask spread was found to be significantly higher under T-1 than under T-3. Applying these tests to compare the result under T-2 with that under T-3, the spread was found to be significantly higher under T-2 than under T-3.

Fig. 1 traces typical results. In Fig. 1 (a), the subjects compete with one another in the early rounds. In Fig. 1 (b), both the subjects offer (30,22).

The last 30 rounds are used to classify the pairs into following five patterns. A pair is classified into C-1 if more than 80% of the winning bid and ask prices in the last 30 rounds indicates (26, 26). A pair is classified into C-2 if more than 80% of the winning bid and ask prices in the last 30 rounds indicate (27, 25). The total ratio between C-1 and C-2 is equal to the ratio of the competitively priced pairs. A pair is classified into C-3 if more than 80% of the winning bid and ask prices in the last 30 rounds belong to the monopolistic

[4] When the total performance was negative, only showing up fee was paid.

Table 1. The Overview under LFT , T-1-k and T-2-k are conducted in order. T-3-i and T-2-i+2 are conducted in order.

	T-1-1	T-1-2	T-3-1	T-3-2	T-3-3
number of pairs	13	12	9	10	13
number of rounds	103	103	103	103	114
max total profit	6337	9954	15960	24744	30044
min total profit	-8444	-14656	385	-909	-5746
average profit per round	28.74	17.19	66.62	64.34	72.93
	T-2-1	T-2-2	T-2-3	T-2-4	T-2-5
number of pairs	13	12	9	10	13
number of rounds	107	107	83	98	101
max total profit	31608	37876	22484	32744	48330
min total profit	2622	3679	3700	4050	3002
average profit per round	98.30	98.60	128.92	110.21	133.17

Table 2. Average Winning Prices and the Bid-Ask Spread in the Last 30 Rounds

	T-1		T-2		T-3	
ask bid-ask	27.02	2.35	29.20	6.50	27.73	3.62
bid spread	24.67		22.70		24.11	

(a) C-2 under T-3

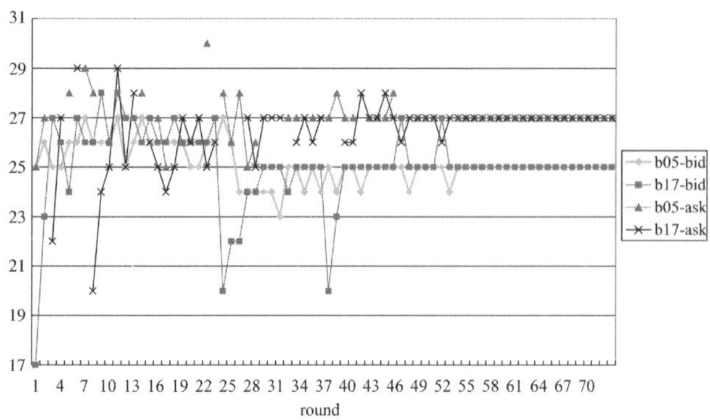

(b) C-4 under T-2

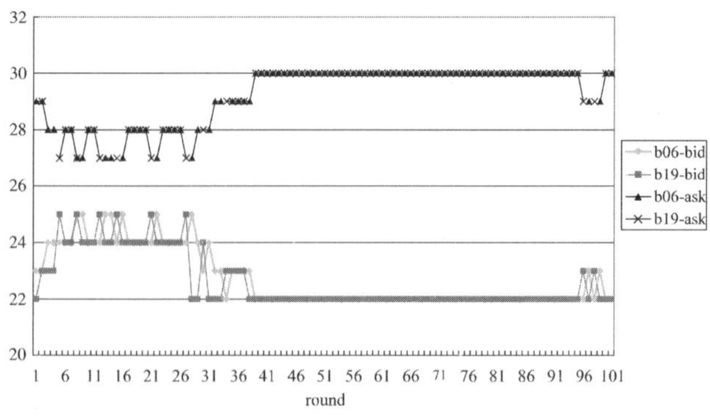

Fig. 1. Typical Results of Convergence

price set. A pair is classified into C-4 if more than 80% of the winning bid and ask prices in the last 30 rounds belong to another price set. In other cases, a pair is classified into C-5.

Table 3. Classification Result under LFT

	C-1(%)	C-2(%)	C-3(%)	C-4(%)	C-5(%)
T-1	0.00	32.00	0.00	4.00	64.00
T-2	1.75	47.37	3.51	7.02	40.35
T-3	0.00	43.75	0.00	0.00	56.25

Table 3 indicates the results of the classification. Comparing T-1 with T-3, the supply-demand information increases the competitiveness. The bid-ask spread difference between T-1 and T-3 is absolutely small but significant, since the C-5 pairs offered more profitable prices under T-3 and half of the C-5 pairs offered the competitive prices under T-1. Comparing T-2 with T-3, the experience effect is significant. As investigated by Huck et al. [3], the result under T-2 in the Table 2 demonstrates the experience makes the subjects move away from the competitive prices. However, the result under the Table 3 shows that the ratio of C-2 is the highest under T-2. This suggests the existence of some competitive factors.

At the end of this section, our results are compared with other experimental results. It is confirmed that our results are the most competitive among the price competition experiments. Dufwenberg and Gneezy [1] and Dufwenberg et al. [2] found that when the number of competitors is two, no subjects offer competitive prices except in the case of price floor treatment, although the experimental settings aim to promote competition (e.g. random matching in every round and informing subjects of the number of rounds). Our results show that 32.00 to 49.12% of the pairs offer market prices and that our ratios are the highest among these experiments, although the experimental settings do not aim at offering the competitive prices (e.g. maintaining identical pairs throughout a treatment and providing the subjects no information about the number of rounds). It is concluded that under all treatments, the convergence rate to the competitive prices is significantly higher than that of Dufwenberg and Gneezy [1] and Dufwenberg et al. [2].

4 Discussion

In this section, the competitive factors in our experiments are examined; subsequently, the relationship between our experiments and the real markets is discussed.

There are two competitive sources: the bid-price competition before sale and the threat of the dead inventory. These factors accelerate the reduction in ask prices and the appreciation in bid prices. First, if our subjects fail to buy a commodity, they must leave the market. Since they hope to participate in the market and beat the opponent, especially in the early rounds, they offer high bid prices. This type of competition is not assumed in the other price competition experiments. Second, the threat of the dead inventory occurs in the second stage. In most price competition experiments, except Plott and Uhl [6], this problem has been ignored since these experiments assume that firms can immediately produce all the units that they wish to sell. On the other hand, we take the dead inventory seriously. If the subjects beat the opponent and buy a commodity but do not sell all the units, they suffer from a significant loss in profit. While they cannot offer high ask prices, but offer ask prices at which they can sell all the units.

Finally, the relation between real markets and our experiments is discussed. The nature of discount ticket shops is the closest to the nature of our middlemen. Those shops in Japan usually buy tickets and gift certificates from suppliers at about 10% discount of the list prices, and then sell them to consumers at about 5% discount of the list prices. The bid-ask spread is extremely small because a number of shops compete with each other and the tickets are homogeneous. Our experimental results suggest this type of stiff competition. As the number of competitors increases, the competitive price is offered more easily. However, our result indicates that the competitive result was attained by at least two middlemen. The most important factors for our experiments are the bid price competition before sale and the dead inventory, rather than the number of competitors.

References

1. Dufwenberg, M. and Gneezy, U., 2000 Price Competition and Market Concentration an Experimental Study, *International Journal of Industrial Organization*, vol 18, pp7-22
2. Dufwenberg M., Gneezy U., Jacob K. G., and Rosemaries N., 2002 Price Floor and Competition. working paper, Stockholm University.
3. Huck, S, Normann, H. T., and Oeshssler, J., 2004 Two are few and four are many: number effects in experimental oligopolies, *Journal of Economic Behavior and Organization*, vol.53, pp.435-446
4. Ogawa, K., Iyori, K. and Oda, S.H., 2004(a) Price Competition Between Middlemen: An Experimental Study, *Gaming, Simulations, and Society: Research Scope and Perspective, Shiratori, Arai, and Kato(edes.)*, pp.59-68,
5. ———, 2004(b) Price Competition between Price Setting Middlemen: in the Laboratory Setting, *Proceedings of EES2004*, pp.176-189
6. Plott, C. R. and Uhl. J. T., 1981 Competitive equilibrium with middlemen: An empirical study, *Southern Economic Journal*, Vol. 47, pp. 1063–71.
7. Spulber, D. F., 1999 *Market Microstructure -Intermediaries and the theory of the firm-*, Cambridge University Press.
8. Stahl, D., O., II, 1988 Bertrand Competition for Inputs and Walrasian Outcomes, *American Economic Review*, vol.78, pp.189-201

Does the Level of Information Matter for Traders? On the Usefulness of Information in Experimental Asset Markets*

Juergen Huber[1], Michael Kirchler[2], and Matthias Sutter[3]

[1] Universitaet Innsbruck
[2] Universitaet Innsbruck
[3] Max-Planck-Institute for Economics Jena

1 Motivation

The literature on the relation between a trader's information about the traded assets and the trader's returns from trading has, so far, mainly concentrated on evaluating the value of insider information. Several papers have shown that insider information leads to returns far above the market average (see, e.g., Jeng et.al. [2], Lin and Howe [5], Lakonishok and Lee [4], Krahnen et.al. [3]). Much less effort has been invested into the question whether the level of information has also a significantly positive impact on the returns from trading when average informed traders are compared to traders with little or almost no information. There is some evidence that managers of small or large investment funds systematically underperform the market (see, e.g., Cowles [1], Malkiel [6][7]), which might be taken as evidence that rather well informed traders (without insider information, though) do not beat the market average. However, whether having no information might lead to better trading performance has not been addressed systematically, so far. In this study we will address the relation between a trader's information level and his returns from trading in an experimental asset market where we can control carefully for a trader's information level about the traded assets. Our results suggest that having more information need not lead to higher returns, except for the very best informed traders.

* We would like to thank Michael Hanke and Klaus Schredelseker for their very helpful comments on earlier versions of this paper. Financial support from the Raiffeisen-Landesbank Tirol and the Center for Experimental Economics at the University of Innsbruck is gratefully acknowledged.

2 The Market Model and Experimental Design

In our market traders can trade stocks of a risky asset or invest at a known risk-free rate. While prices are set solely by the actions of the traders, the risky asset's intrinsic value is determined by a random walk dividend stream:

$$D_t = D_{t-1} + e \tag{1}$$

D_t denotes the dividend in period t, D_0 is set to 0.2, and e is a normally distributed random term with a mean of zero and a variance of 0.0004. The sequence of dividends has been randomly determined before running the experimental sessions and has been applied to in all experimental sessions in order to guarantee identical conditions. In one experimental market, there are 9 traders denoted $I1$ to $I9$. The label Ix indicates that a trader knows the dividend of this and the next $(x-1)$ periods. For instance, trader $I5$ knows the asset's dividends of this and the next 4 periods. For the sake of simplicity we assume that traders know the exact value of the future dividends. At the end of each period the current dividend is paid out for each stock owned. In the next period the information on dividends is updated, such that the former dividend for period t+1 is now the dividend of period t. All traders in a group started with an endowment of 1600 units of cash (Taler) and 40 stocks. Then they were informed about the future dividends according to their information level, and trading started. Trading time was 100 seconds per period and in total, we had 30 periods. In addition to his information level, we displayed to each trader the net present value of the stock, by using Gordon's formula, i.e. discounting the known dividends and assuming the last one to be a continuous, infinite stream which was also discounted.[1]

$$E(V, Ij_t) = \frac{D_{t+j-1}}{(1+r_e)^{j-2} * r_e} + \sum_{i=t}^{t+j-2} \frac{D_i}{(1+r_e)^{i-t}} \tag{2}$$

$E(V, Ij_t)$ denotes the conditional expected value of the asset in period t for information level Ij. The risk adjusted interest rate r_e was fixed at 0.5%. The expected growth rate of the dividend was set to zero and is therefore not shown in the formula. The resulting paths of the conditional expected values of the asset for uneven information levels for k=30 periods in our experimental treatment are shown in Figure 1.[2] Compared to information level $I9$ the expected values known to a trader with information level Ij are shifted $(9-j)$ periods to the right. This means that better informed traders receive information earlier than less informed traders, which is a key feature in a market with asymmetrically informed traders. The experimental sessions were implemented with a total of 7 independent groups with 9 subjects each.

[1] Subjects were informed in the instructions how the net present value was calculated and that it depended on the information level.

[2] For the sake of clarity we only plotted the uneven information levels.

Each of the 9 subjects had a different information level ranging from $I1$ to $I9$. The average duration of the sessions was 90 minutes, with average earnings of 18 Euro.

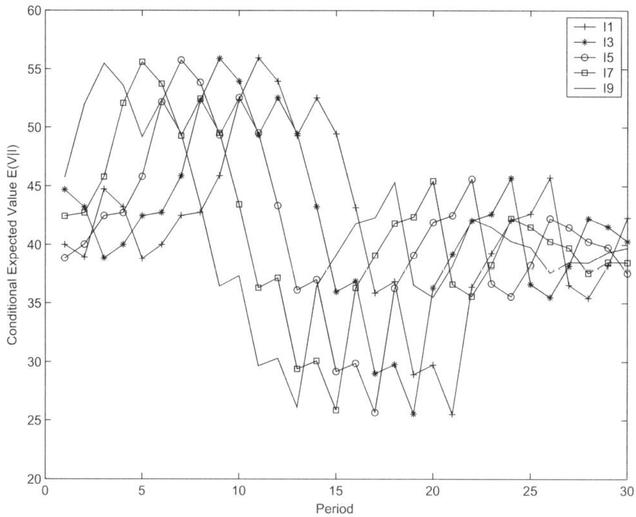

Fig. 1. Conditional expected values as a function of period of the uneven information levels

3 Experimental Results

Figure 2 shows the individual and average returns (solid line) for different information levels. The returns are calculated according to the following formula (3), where $R_{t,0,j}$ denotes the return of trader Ij from period t (with wealth $W_{t,j}$) to period T (with wealth $W_{0,j}$). Setting t to 30 we compare a trader's wealth at the end with the one at the beginning of the experiment.

$$R_{t,0,j} = \frac{W_{t,j} - W_{0,j}}{W_{0,j}} \tag{3}$$

We find that the best informed traders earn on average the highest returns, but in general there is no positive relationship between a trader's information level and his return. Average returns range from 7.1% for traders $I5$ to 22.2% for traders $I9$. Due to a relatively high variance in single traders' returns a Friedman test shows that there is no significant difference in the returns of traders with different information levels ($p = 0.11$; two-sided Friedman test including all information levels, $N = 7$). Only when we check pairwise differences, we find that the average returns are significantly higher for traders

*I*9 than for traders *I*3, *I*4 or *I*5 ($p<0.05$ in each pair wise test, Wilcoxon signed ranks test, N = 7). Overall, 22 out of 63 traders exceeded the expected return of the stock of 16.1% (0.5% for 30 periods = 1.005^{30}-1). 9 out of 63 traders finished with a return which was lower than the risk-free rate of approximately 3.0% (0.1% for 30 periods = 1.001^{30}-1), with most of them actually making losses. In sum, we can say that there is a broad range of information levels where additional information has no significantly positive influence on returns, and that only the very best informed traders can actually outperform (some of) the less informed ones.[3]

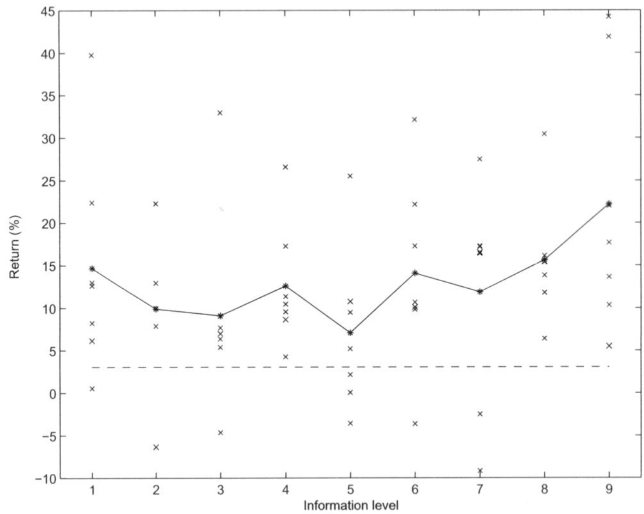

Fig. 2. Individual and average returns in percent per information level

The source for the non-monotonicity between information levels and returns from trading can be detected if we look at the conditional expected values per information level and at the endogenously evolving stock holdings of each information level.[4] The best informed trader *I*9 is the first one to start buying actively (and at relatively low prices), because he is the first one realizing that dividends are steadily increasing from period 5 to period 12 (see the solid line in the left panel of Figure 3). In addition, when prices are high and prospects for dividends deteriorating, trader *I*9 is the first to sell. Traders *I*5 who have the lowest average returns are the most eager ones to sell stocks

[3] Note, that two outliers with 53.25 (*I*9) and -30.65 (*I*2) are not plotted in the graph. The dotted line represents the risk free interest rate of 3 percent for 30 periods.

[4] For the sake of clarity we have selected only the uneven information levels in Figure 3. A separate figure with the stock holdings of traders with an even information level shows a very similar pattern and is available upon request.

at the beginning of the experiment, as they have the lowest estimates of the present value. These stocks are quite frequently bought by traders $I9$. When traders $I5$ realize around period 6 that dividends increase in the future they start buying stocks actively. As average prices are leading dividends by 1 to 7 periods, the peak of prices can be observed earlier than the peak of dividends in period 11 and so average informed traders buy at the peak of the price path around period 8 as can be seen from Figure 3.

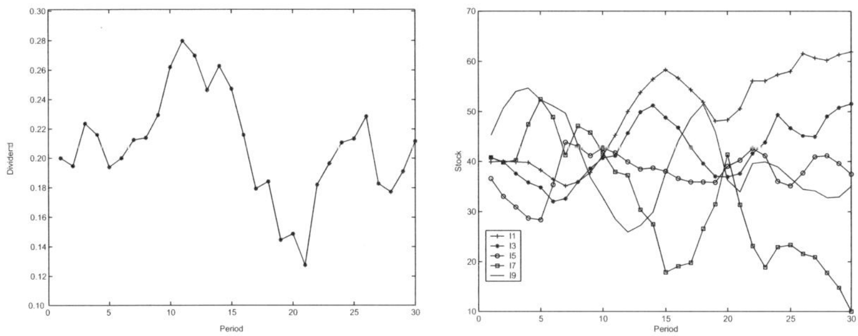

Fig. 3. Left: Dividends across periods, Right: Average stock holdings per information level

So far, we have only considered the final wealth of subjects and their returns from trading over all 30 periods. Yet, it might be interesting to check whether information levels and returns are somewhat differently related in earlier periods of the experiment. To do so, we compare the average wealth per information level with the initial wealth $W_{0,j}$ of information level j. From that we can calculate the average return $R_{t,0,j}$ according to equation (3). The results are displayed in Figure 4.

We see that the performance of traders across time is remarkably stable. The insiders $I8$ and $I9$ have the best performance at the beginning, in the middle, and at the end of the experiment. Similarly, the finally worst performing traders $I5$ and $I3$ are already lagging behind after the first few periods. In general, there are very few intersections in Figure 4. Rather the differences increase over time. The distribution of final average returns is therefore not due to a few periods, but it is the result of different performance throughout the experiment.

4 Conclusion

In this study we have shown that more information about a tradable asset need not lead to higher returns from trading this asset. Although the very best informed traders outperform most of the others, there is no statistically significant difference between the returns of medium informed traders and

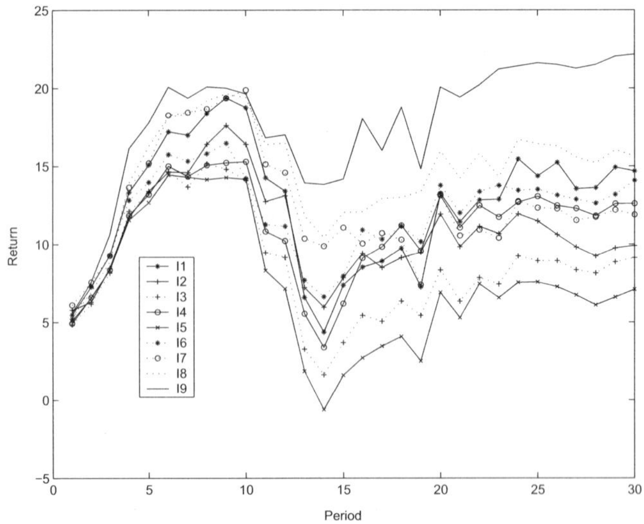

Fig. 4. Development of average returns $R_{t,0,j}$ over time

those with almost no information. It seems to be a worthwhile topic for future research to explore the reasons for this pattern and the stability of this pattern in more depth.

References

1. Cowles, A. (1933), Can Market Forecasters Forecast?, Econometrica, Vol. 1, p. 309-324.
2. Jeng, L.A., Metrick, A., and Zeckhauser, R. (2003), Estimating the Returns to Insider Trading: A Performance-Evaluation Perspective, The Review of Economics and Statistics, Vol. 85, p. 453-471.
3. Krahnen, J.P., Rieck, C., and Theissen, E. (1999), Insider trading and portfolio structure in experimental asset markets with a long-lived asset, European Journal of Finance, Vol. 5, p. 29-50.
4. Lakonishok, J. and Lee, I. (2001), Are insider trades informative?, The Review of Financial Studies, Vol. 14, p. 79-111.
5. Lin, J. and Howe, J. (1990), Insider trading in the OTC market, Journal of Finance, Vol. 45, p. 1273-1284.
6. Malkiel, B.G. (2003a), Passive Investment Strategies and Efficient Markets, European Financial Management, Vol. 9 (1), p. 1-10.
7. Malkiel, B.G. (2003b), The Efficient Market Hypothesis and Its Critics, Journal of Economic Perspectives, Vol. 17 (1), p. 59-82.

A Laboratory Comparison of Arbitration Mechanisms: FOA and AFOA

Cary Deck[1], Amy Farmer[2], and Dao-Zhi Zeng[3]

[1] University of Arkansas
[2] University of Arkansas
[3] Kagawa University

1 Introduction

Given its lower costs relative to litigation, arbitration is rapidly increasing as a mechanism to settle disputes. While it has traditionally been employed in labor disputes, arbitration has several characteristics that are making it an increasingly attractive option in numerous forms of dispute. First, relative to traditional litigation, arbitration is significantly less costly. However, in addition to the lower expense, arbitration generally places greater restrictions on discovery thereby lessening the competitive information that a business may have to reveal in the resolution of a dispute. Further, these limitations tend to hasten the process further reducing both direct and indirect costs. As mentioned by Fuller [9], Apple Computer saved over $4 million in legal fees by using arbitration in a case with the IRS and simultaneously prevented the revelation of proprietary information which would have occurred in standard litigation. As a result of these characteristics, arbitration has become the dispute resolution of choice for numerous business transactions.

Given that the use of arbitration to settle disputes is becoming increasingly more widespread, it is important that scholars examine the rules and procedures associated with various forms of arbitration available to disputants in order to design a system that is the most efficient for parties involved. It is widely understood that the form of arbitration can generate different incentives for strategic bidding by the parties to the dispute. The most commonly used forms are conventional arbitration (CA) in which the arbitrator is free to make the ruling as he or she sees fit and final-offer arbitration (FOA) proposed by Stevens [13] in which an arbitrator must choose between final offers submitted by each party, i.e., there is no splitting the difference.

While it is generally agreed that the most efficient system is the one that produces the greatest degree of self-negotiated settlement, law and economics scholars have only begun to understand the variety of bargaining incentives generated from alternative forms of arbitration, and as yet, there is no consensus as to the most efficient form. Theory suggests that when the disputants have symmetric information, any form of arbitration should generate settlement within a contract zone which defines the set of Pareto improving settlements. This contract zone always exists as long as arbitration involves some costs, and theory predicts that disputants will settle 100% of the time within this region, arbitration can be avoided and disputes are settled efficiently. Unfortunately, neither CA nor FOA has been successful at eliminating disputes in practice. As with standard litigation, the question why there is ever disagreement in cases where agreement appears to be in the interests of both parties is an enduring puzzle which researchers have attempted to address. The predominant theoretical explanation is that parties may hold asymmetric information regarding the outcome, thereby preventing the ability to reach agreement within this contract zone. This asymmetry can lead to strategic screening and signaling behaviors that are rational and optimal ex ante, but lead to the inefficient use of arbitration or litigation ex post. See Farmer and Pecorino [8] for a survey of asymmetric information in bargaining.

2 Amended Final-Offer Arbitration: Theory

However, in laboratory experiments in which information can be controlled and constructed to be symmetric, both CA and FOA have failed to generate settlement as predicted (see Ashenfleter et al [1], Deck and Farmer [4, 5], and Dickinson [6]). Laboratory results typically find that both FOA and CA generate settlement rates of about 50%, which is less than what is observed in the absence of any arbitration method. More importantly perhaps, there seems to be a consensus that CA outperforms FOA in terms of settlement rates.

Given the inability of either mechanism to produce the high level of settlement predicted by theory even in a controlled environment with symmetric information, researchers have developed various alternative mechanisms. Other forms of arbitration which are variations of CA or FOA include tri-offer (Ashenfleter et al [1]) and combined (Brams and Merill [3]). Tri-offer arbitration adds a third party who makes a recommendation at an earlier stage in the negotiation process, and as such, the arbitrator is free to choose among the offers of the two disputants or the offer from the third neutral party. Combined arbitration is a mechanism in which FOA rules are used when the arbitrator's value lies between the two offers, and CA is implemented when it lies on either side. Ashenfelter et al [1] and Dickinson [6] test tri-offer and combined arbitration in the laboratory respectively and both find that these alternative mechanisms perform worse in terms of settlement rates.

In the wake of those works and the failings of researchers to design a mechanism that improves upon either CA or FOA when it is implemented in the laboratory, Zeng [14] proposed yet another alternative dispute resolution mechanism, referred to as amended final-offer arbitration (AFOA), which represents a simple amendment to FOA. Under AFOA, the individual's bid affects the probability that he or she wins, but the award amount is determined by the deviation of the opponent's bid from the arbitrator's value. Specifically, as with FOA, the arbitrator chooses the winner based upon whose bid is closer to his or her preferred value, but the payout equals the arbitrator's value plus an adjustment which is based upon how far the losing party's bid was from the arbitrator's value. Thus, the winning player's ultimate payout is independent of the bid submitted. By separating the probability of winning from the payment, AFOA provides the incentive to bid reasonably in order to increase the chance of winning and simultaneously prevent an even more unfavorable outcome in the event that one's bid loses. In so doing, a player does not harm his eventual payout if he is deemed the winner. As a result, Zeng [14] demonstrates theoretically that offers converge in equilibrium in AFOA. In contrast, Farber [7] and Brams and Merrill [2] find that the opposite is true of equilibrium bidding in FOA. Specifically, disputants want to make reasonable offers in order to win, but in doing so compromise their payout. As a result, the equilibrium bidding behavior results in divergent offers, a difficulty that AFOA does not have. While theory predicts that settlement should occur regardless of bidding behavior due to the presence of a contract zone, it is clear that this is not the reality. It is thought that convergent offers eliminate the uncertainty of pursuing arbitration making the arbitration outcome apparent to both parties ex ante, thereby increasing the probability that settlement will take place in practice.

Despite its attractive theoretical properties and potential social benefits, AFOA is new and untested by both scholars and practitioners alike. It is understandable that disputants in the naturally occurring economy would be reluctant to try an unproven mechanism. And, as with tri-offer and combined arbitration detailed above, new mechanisms that are theoretically attractive are not necessarily functionally successful. As detailed by Smith [12] (p. 115), one of the reasons for conducting experiments is to use "The laboratory as a testing ground for institutional design." The purpose of this paper is to fill the gap and report a series of laboratory experiments on AFOA.

Furthermore, for a more robust evaluation of AFOA this study alternatively models the arbitrator as having a continuous distribution of preferred outcomes and a binary distribution of preferred outcomes. All of the work mentioned above considers only a continuous arbitrator distribution, and in many real world situations, modeling the arbitrator preference distribution as continuous is appropriate. For example, in terms of award cash payment for damages or settling a salary dispute, a continuous distribution would be preferred. But in some cases a model using a binary distribution is better. For example, there might be two types of arbitrators, one likely to rule one way and another likely to rule the other. Or, in a custody battle between parents, a parent who is seeking custody of a child will either be successful or not. As arbitration usage is expanding beyond its traditional applications, these situations are certainly relevant. However, most of the literature including all of the experimental work has focused on the continuous case only. Interestingly,

Kilgour [11] analyzed FOA using a binary distribution and found that optimal behavior changes significantly; specifically, when the arbitrator's distribution is binary, no pure strategy equilibrium exists. Another strength of AFOA is that theoretically it is robust to such changes, yet another attractive property of AFOA from a theoretical perspective. In this paper we examine in the laboratory whether AFOA performs as theory predicts using both a continuous distribution for arbitrator preferences as well as a binary distribution. As such, we simultaneously examine FOA in both scenarios, thereby testing FOA behavior in the case of a binary distribution which has to date not been studied.

3 Experimental Design and Results

We conducted four replicates in each cell of a 2×2 design, where the first dimension refers to the arbitration mechanism (AFOA vs FOA) and the second dimension is the arbitrator's distribution (uniform vs binary). Each of our experimental sessions is composed of two phases. In the first phase, subjects are not allowed to bargain and instead are forced to enter arbitration. As such, we were able to observe bargaining behavior without having to rely on bargaining failure. This first phase allowed us to design experiments without imposing conditions that might contribute to bargaining failure as previous experiments have done. For example, Ashenfelter et al [1] used an normal distribution and then guided subjects to bargain over a range that was not centered at the mean. This was intended to prevent subjects from simply settling at the equal split. Dickinson [6] effectively created asymmetric information by presenting differing bargaining ranges to the subjects, again preventing the possibility that subjects simply resort to an equal split. Finally, Deck and Farmer [4] do not impose asymmetric information, but the form of the symmetric uncertainty was sufficiently complex that some bargaining failure was expected. Specifically, given the form of the uncertainty, the distribution was asymmetric and, once again, the design guided subjects away from reverting automatically to the equal split. In each of those previous papers, the experiment was designed in order to prevent the possibility of high rates of settlement thereby preventing the researchers from observing bidding behavior.

While those works demonstrate the relative failure of FOA to CA, they necessitate a complexity in the experiment design that effectively induces settlement failure. In this paper we are able to prevent that problem in that our two phase design does not require settlement failure in order for us to observe bidding behavior. In the second phase of the experiment, subjects were allowed to bargain with each other, with arbitration being used only when a settlement was not reached. Given that they have already experienced arbitration in the first phase of the experiment, this ordering provides the additional benefit that subjects then have the opportunity to more fully understand the workings of arbitration before they are asked to engage in pre-arbitration negotiations.

Based upon 1280 individual allocation decisions that we ultimately observe, the experimental results confirm the theoretical merits of AFOA, indicating that AFOA outperforms FOA in several important ways. First, the offers that disputants place in AFOA conform to the theoretical predictions and tend to converge while those produced by FOA disputants do not. Second, AFOA leads to greater pre-arbitration agreement than does FOA, a result consistent with the hypothesis that convergent bids will reduce the uncertainty surrounding arbitration and create a focal point for settlement. Specifically, settlement rates are 93% in AFOA as compared to only 75% in FOA. Finally, AFOA tends to conform to its theoretical predictions more closely than does FOA. Specifically, offers converge in AFOA as predicted, but they do not diverge in FOA as predicted. In fact, in FOA the offers are neither convergent nor divergent. Thus, in addition to the positive findings that AFOA produces convergent offers and greater settlement, the simple fact that AFOA as practiced in the laboratory conforms to theory suggests that it is a more predictable and stable mechanism that is worthy of greater attention from both scholars and practitioners alike.

4 Concluding Remarks

As indicated in Zeng [14], the basic idea of AFOA comes from that of a second-price auction. While theory predicts truthful-revelation in the second-price auction, laboratory behavior does not conform to this prediction (see Harstad [10]). Therefore, the success of AFOA in its implementation is not only useful but is also somewhat surprising. Harstad [10] explains the inconsistency of second price auctions in the laboratory as a result of the existence of a positive feedback; i.e., a bidder may still win and earn a positive payoff even when overbidding if the bidder would have won with truthful revelation. This positive feedback is similar to the contrasting incentives present in FOA. However, in AFOA this positive feedback is eliminated because an overbid strategy generates an unambiguous loss for the disputant.

In sum, Amended Final-Offer Arbitration shows tremendous promise for generating predictable outcomes in line with theory, eliminating the uncertainty regarding arbitration, thereby encouraging settlement and increasing efficiency. Due to the significant potential social benefits, the results of this study indicate that further analysis of this mechanism is clearly warranted.

References

1. Ashenfelter O, Currie J, Farber HS, Spiegel M (1992) An experimental comparison of dispute rates in alternative arbitration systems. Econometrica 60:1407–1433

2. Brams SJ, Merrill SIII (1983) Equilibrium strategies for final-offer arbitration: there is no median convergence. Management Science 29:927–941
3. Brams SJ, Merrill SIII (1986) Binding versus final-offer arbitration: a combination is best. Management Science 32:1346–1355
4. Deck C, Farmer A (2005a) Bargaining over an uncertain value: arbitration mechanisms compared. Journal of Law, Economics and Organization, forthcoming.
5. Deck C, Farmer A (2005b) Bargaining with asymmetric uncertainty: arbitration mechanisms compared. Working paper, University of Arkansas
6. Dickinson DL (2004) A comparison of conventional, final-offer, and "combined" arbitration for dispute resolution. Industrial and Labor Relations Review 57:288–301
7. Farber HS (1980) An analysis of final-offer arbitration. Journal of Conflict Resolution 24:683–705
8. Farmer A, Pecorino P (1996) Issues of information asymmetry in legal bargaining. In: Anderson D (ed) Dispute resolution: bridging the settlement gap, JAI Press, Inc.
9. Fuller, JP (1993) U.S. Tax Review. Tax Notes International 7:1041–1042.
10. Harstad RM (2000) Dominant strategy adoption and bidders' experience with pricing rules. Experimental Economics 3:261–280
11. Kilgour DM (1994) Game-theoretic properties of final-offer arbitration. Group Decision and Negotiation 3:285–301
12. Smith VL (1994) Economics in the laboratory. Journal of Economic Perspectives 8:113–131
13. Stevens CM (1966) Is compulsory arbitration compatible with bargaining? Industrial Relations 5:38–52
14. Zeng D-Z (2003) An amendment to final-offer arbitration. Mathematical Social Sciences 46:9–19

Lecture Notes in Economics and Mathematical Systems

For information about Vols. 1–505
please contact your bookseller or Springer-Verlag

Vol. 506: B. P. Kellerhals, Financial Pricing Models in Continuous Time and Kalman Filtering. XIV, 247 pages. 2001.

Vol. 507: M. Koksalan, S. Zionts, Multiple Criteria Decision Making in the New Millenium. XII, 481 pages. 2001.

Vol. 508: K. Neumann, C. Schwindt, J. Zimmermann, Project Scheduling with Time Windows and Scarce Resources. XI, 335 pages. 2002.

Vol. 509: D. Hornung, Investment, R&D, and Long-Run Growth. XVI, 194 pages. 2002.

Vol. 510: A. S. Tangian, Constructing and Applying Objective Functions. XII, 582 pages. 2002.

Vol. 511: M. Külpmann, Stock Market Overreaction and Fundamental Valuation. IX, 198 pages. 2002.

Vol. 512: W.-B. Zhang, An Economic Theory of Cities.XI, 220 pages. 2002.

Vol. 513: K. Marti, Stochastic Optimization Techniques. VIII, 364 pages. 2002.

Vol. 514: S. Wang, Y. Xia, Portfolio and Asset Pricing. XII, 200 pages. 2002.

Vol. 515: G. Heisig, Planning Stability in Material Requirements Planning System. XII, 264 pages. 2002.

Vol. 516: B. Schmid, Pricing Credit Linked Financial Instruments. X, 246 pages. 2002.

Vol. 517: H. I. Meinhardt, Cooperative Decision Making in Common Pool Situations. VIII, 205 pages. 2002.

Vol. 518: S. Napel, Bilateral Bargaining. VIII, 188 pages. 2002.

Vol. 519: A. Klose, G. Speranza, L. N. Van Wassenhove (Eds.), Quantitative Approaches to Distribution Logistics and Supply Chain Management. XIII, 421 pages. 2002.

Vol. 520: B. Glaser, Efficiency versus Sustainability in Dynamic Decision Making. IX, 252 pages. 2002.

Vol. 521: R. Cowan, N. Jonard (Eds.), Heterogenous Agents, Interactions and Economic Performance. XIV, 339 pages. 2003.

Vol. 522: C. Neff, Corporate Finance, Innovation, and Strategic Competition. IX, 218 pages. 2003.

Vol. 523: W.-B. Zhang, A Theory of Interregional Dynamics. XI, 231 pages. 2003.

Vol. 524: M. Frölich, Programme Evaluation and Treatment Choise. VIII, 191 pages. 2003.

Vol. 525: S. Spinler, Capacity Reservation for Capital- Intensive Technologies. XVI, 139 pages. 2003.

Vol. 526: C. F. Daganzo, A Theory of Supply Chains. VIII, 123 pages. 2003.

Vol. 527: C. E. Metz, Information Dissemination in Currency Crises. XI, 231 pages. 2003.

Vol. 528: R. Stolletz, Performance Analysis and Optimization of Inbound Call Centers. X, 219 pages. 2003.

Vol. 529: W. Krabs, S. W. Pickl, Analysis, Controllability and Optimization of Time-Discrete Systems and Dynamical Games. XII, 187 pages. 2003.

Vol. 530: R. Wapler, Unemployment, Market Structure and Growth. XXVII, 207 pages. 2003.

Vol. 531: M. Gallegati, A. Kirman, M. Marsili (Eds.), The Complex Dynamics of Economic Interaction. XV, 402 pages, 2004.

Vol. 532: K. Marti, Y. Ermoliev, G. Pflug (Eds.), Dynamic Stochastic Optimization. VIII, 336 pages. 2004.

Vol. 533: G. Dudek, Collaborative Planning in Supply Chains. X, 234 pages. 2004.

Vol. 534: M. Runkel, Environmental and Resource Policy for Consumer Durables. X, 197 pages. 2004.

Vol. 535: X. Gandibleux, M. Sevaux, K. Sörensen, V. T'kindt (Eds.), Metaheuristics for Multiobjective Optimisation. IX, 249 pages. 2004.

Vol. 536: R. Brüggemann, Model Reduction Methods for Vector Autoregressive Processes. X, 218 pages. 2004.

Vol. 537: A. Esser, Pricing in (In)Complete Markets. XI, 122 pages, 2004.

Vol. 538: S. Kokot, The Econometrics of Sequential Trade Models. XI, 193 pages. 2004.

Vol. 539: N. Hautsch, Modelling Irregularly Spaced Financial Data. XII, 291 pages. 2004.

Vol. 540: H. Kraft, Optimal Portfolios with Stochastic Interest Rates and Defaultable Assets. X, 173 pages. 2004.

Vol. 541: G.-y. Chen, X. Huang, X. Yang, Vector Optimization. X, 306 pages. 2005.

Vol. 542: J. Lingens, Union Wage Bargaining and Economic Growth. XIII, 199 pages. 2004.

Vol. 543: C. Benkert, Default Risk in Bond and Credit Derivatives Markets. IX, 135 pages. 2004.

Vol. 544: B. Fleischmann, A. Klose, Distribution Logistics. X, 284 pages. 2004.

Vol. 545: R. Hafner, Stochastic Implied Volatility. XI, 229 pages. 2004.

Vol. 546: D. Quadt, Lot-Sizing and Scheduling for Flexible Flow Lines. XVIII, 227 pages. 2004.

Vol. 547: M. Wildi, Signal Extraction. XI, 279 pages. 2005.

Vol. 548: D. Kuhn, Generalized Bounds for Convex Multistage Stochastic Programs. XI, 190 pages. 2005.

Vol. 549: G. N. Krieg, Kanban-Controlled Manufacturing Systems. IX, 236 pages. 2005.

Vol. 550: T. Lux, S. Reitz, E. Samanidou, Nonlinear Dynamics and Heterogeneous Interacting Agents. XIII, 327 pages. 2005.

Vol. 551: J. Leskow, M. Puchet Anyul, L. F. Punzo, New Tools of Economic Dynamics. XIX, 392 pages. 2005.

Vol. 552: C. Suerie, Time Continuity in Discrete Time Models. XVIII, 229 pages. 2005.

Vol. 553: B. Mönch, Strategic Trading in Illiquid Markets. XIII, 116 pages. 2005.

Vol. 554: R. Foellmi, Consumption Structure and Macroeconomics. IX, 152 pages. 2005.

Vol. 555: J. Wenzelburger, Learning in Economic Systems with Expectations Feedback (planned) 2005.

Vol. 556: R. Branzei, D. Dimitrov, S. Tijs, Models in Cooperative Game Theory. VIII, 135 pages. 2005.

Vol. 557: S. Barbaro, Equity and Efficiency Considerations of Public Higer Education. XII, 128 pages. 2005.

Vol. 558: M. Faliva, M. G. Zoia, Topics in Dynamic Model Analysis. X, 144 pages. 2005.

Vol. 559: M. Schulmerich, Real Options Valuation. XVI, 357 pages. 2005.

Vol. 560: A. von Schemde, Index and Stability in Bimatrix Games. X, 151 pages. 2005.

Vol. 561: H. Bobzin, Principles of Network Economics. XX, 390 pages. 2006.

Vol. 562: T. Langenberg, Standardization and Expectations. IX, 132 pages. 2006.

Vol. 563: A. Seeger (Ed.), Recent Advances in Optimization. XI, 455 pages. 2006.

Vol. 564: P. Mathieu, B. Beaufils, O. Brandouy (Eds.), Artificial Economics. XIII, 237 pages. 2005.

Vol. 565: W. Lemke, Term Structure Modeling and Estimation in a State Space Framework. IX, 224 pages. 2006.

Vol. 566: M. Genser, A Structural Framework for the Pricing of Corporate Securities. XIX, 176 pages. 2006.

Vol. 567: A. Namatame, T. Kaizouji, Y. Aruga (Eds.), The Complex Networks of Economic Interactions. XI, 343 pages. 2006.

Vol. 568: M. Caliendo, Microeconometric Evaluation of Labour Market Policies. XVII, 258 pages. 2006.

Vol. 569: L. Neubecker, Strategic Competition in Oligopolies with Fluctuating Demand. IX, 233 pages. 2006.

Vol. 570: J. Woo, The Political Economy of Fiscal Policy. X, 169 pages. 2006.

Vol. 571: T. Herwig, Market-Conform Valuation of Options. VIII, 104 pages. 2006.

Vol. 572: M. F. Jäkel, Pensionomics. XII, 316 pages. 2006

Vol. 573: J. Emami Namini, International Trade and Multinational Activity. X, 159 pages, 2006.

Vol. 574: R. Kleber, Dynamic Inventory Management in Reverse Logistics. XII, 181 pages, 2006.

Vol. 575: R. Hellermann, Capacity Options for Revenue Management. XV, 199 pages, 2006.

Vol. 576: J. Zajac, Economics Dynamics, Information and Equilibnum. X, 284 pages, 2006.

Vol. 577: K. Rudolph, Bargaining Power Effects in Financial Contracting. XVIII, 330 pages, 2006.

Vol. 578: J. Kühn, Optimal Risk-Return Trade-Offs of Commercial Banks. IX, 149 pages, 2006.

Vol. 579: D. Sondermann, Introduction to Stochastic Calculus for Finance. X, 136 pages, 2006.

Vol. 580: S. Seifert, Posted Price Offers in Internet Auction Markets. IX, 186 pages, 2006.

Vol. 581: K. Marti; Y. Ermoliev; M. Makowsk; G. Pflug (Eds.), Coping with Uncertainty. XIII, 330 pages, 2006.

Vol. 582: J. Andritzky, Sovereign Default Risks Valuation: Implications of Debt Crises and Bond Restructurings. VIII, 251 pages, 2006.

Vol. 583: I.V. Konnov, D.T. Luc, A.M. Rubinov† (Eds.), Generalized Convexity and Related Topics. IX, 469 pages, 2006.

Vol. 584: C. Bruun, Adances in Artificial Economics: The Economy as a Complex Dynamic System. XVI, 296 pages, 2006.

Vol. 585: R. Pope, J. Leitner, U. Leopold-Wildburger, The Knowledge Ahead Approach to Risk. XVI, 218 pages, 2007 (planned).

Vol. 586: B.Lebreton, Strategic Closed-Loop Supply Chain Management. X, 150 pages, 2007 (planned).

Vol. 587: P. N. Baecker, Real Options and Intellectual Property: Capital Budgeting Under Imperfect Patent Protection. X, 276 pages , 2007.

Vol. 588: D. Grundel, R. Murphey, P. Panos , O. Prokopyev (Eds.), Cooperative Systems: Control and Optimization. IX, 401 pages , 2007.

Vol. 589: M. Schwind, Dynamic Pricing and Automated Resource Allocation for Information Services: Reinforcement Learning and Combinatorial Auctions. XII, 293 pages , 2007.

Vol. 590: S. H. Oda, Developments on Experimental Economics: New Approaches to Solving Real-World Problems. XVI, 262 pages, 2007.

Vol. 591: M. Lehmann-Waffenschmidt, Economic Evolution and Equilibrium: Bridging the Gap. VIII, 272 pages, 2007.

Vol. 592: A. C.-L. Chian, Complex Systems Approach to Economic Dynamics. X, 95 pages, 2007.

Vol. 593: J. Rubart, The Employment Effects of Technological Change: Heterogenous Labor, Wage Inequality and Unemployment. XII, 209 pages, 2007

Vol. 594: R. Hübner, Strategic Supply Chain Management in Process Industries: An Application to Specialty Chemicals Production Network Design. XII, 243 pages, 2007

Vol. 595: H. Gimpel, Preferences in Negotiations: The Attachment Effect. XIV, 268 pages, 2007

Vol. 596: M. Müller-Bungart, Revenue Management with Flexible Products: Models and Methods for the Broadcasting Industry.XXI, 297 pages, 2007